USEFUL BUSH PLANTS

USEFUL BUSH PLANTS

PETER BINDON

Western Australian Museum

Reprinted 1998

The National Library of Australia
Cataloguing-in-Publication entry:

Bindon, Peter.
Useful Bush Plants.

Includes index.
ISBN 0 7309 7024 8.

1. Plants, Useful – Australia. 2. Plants, Edible – Australia.
3. Native plants for cultivation – Australia.
I. Western Australian Museum. II. Title.

635.95194

Edited by Kate Hooper

Published by the Western Australian Museum
Francis Street, Perth, Western Australia 6000

Cover designed by Robyn Mundy
Printed in Western Australia by Lamb Printers Pty Ltd

Front cover: Background: Blackboy, *Xanthorrhoea preissii*
Inset: Rock Fig, *Ficus platypoda*
Frontispiece: Round-fruited Banksia, *Banksia sphaerocarpa*

FOREWORD

I first encountered Peter Bindon when he was talking about useful bush plants at a nursery industry conference and have become a great fan!

We have worked together on a couple of projects related to bush tucker and I'm constantly impressed by the depth of his knowledge.

In Australia, we're on the brink of a whole new era of interest, in the incredible range of bush plants we are blessed with – and that are the subject of this fascinating book.

In the last decade of the 20th century we're already seeing major restaurants specialising in bush tucker, small cafes where you can enjoy a cappuccino flavoured with wattle seed, and a whole range of bush condiments. But in the 21st century, this trend is going to take off!

I can see useful varieties of bush plants being stocked in nurseries and being cultivated in people's gardens to a far greater extent. Not only do these plants look good, but they have stood the test of time. The wisdom that acknowledged their value - as food or medicine - is as relevant today as it was in ancient times.

Early favourites are likely to be the riberry, a variety of lilly pilly, because it's very pretty and very delicious. One of my favourites is a little *midyim* myrtle which grows naturally around Morton Bay in Queensland. It has abundant and attractive tiny fruit, but it also makes a great ground cover. Then there is the bullrush, a really surprising plant – it's new leaf stalks taste like baby sweet corn, and its starchy roots can be cooked as a delicious damper. And there's the bunya bunya pine, which produces a delicious nut, not unlike the chestnut.

The whole exciting process of rediscovering and cultivating plants that once thrived in an area of pristine bushland near you is something that is happening not just in Australia, but all over the world. And it makes good sense. Regardless of where you live, local bush plants will grow well in your garden because such plants don't require any additional water once established. You help them settle in and you forget about them! And once you have restored these plants to the landscape, they in turn bring back the native birds and animals, some of which are entirely dependant on just a few species for pollen supplies or nectar.

It's early days yet for this fledgling industry, but nursery stocks will grow – because the interest is there, and so is the potential, for both local and international sales.

This book is a great starting point to whet your appetite, and it may well launch you on a journey of discovery in relation to our native plants. I have no hestitation in recommending it to you.

NEVILLE PASSMORE

Neville Passmore is a well-known nurseryman who has established a wide following as the Western Australian representative on the national ABC Television programme "Gardening Australia". A contributor to Radio 6PR's gardening talk-back and gardening columnist for the *Sunday Times*, he works with the Waldeck Group, and is the author of several books including *Roses WA*.

CONTENTS

NOTES ON ENTRIES

Plant families and authors are as for citations in: Hnatiuk, R.J., 1992 *Census of Australian Vascular Plants.* Australian Fauna and Flora Series No.11, Australian Government Publishing Service, Canberra.

A number of species formerly placed in the genus *Eucalyptus* have been reassigned to a new genus *Corymbia*, however they are listed here under their earlier names.

Australia-wide distributions for the plants in the main part of the book were not always available. To maintain formatting consistency, some maps indicate with a question mark the occurrence of the plant outside Western Australia.

ACKNOWLEDGMENTS

Without the assistance of all the people who have discussed plants with me over the years, this book would have been impossible. Many assisted in special ways, and it would take several pages to give details of the ways they helped. I thank: Ray Aitken, Sean Brennan, Bob Cooper, Bob Currie, John Dell, Kingsley Dixon, Emily Dortch, Paul Foulkes, Moira Geoghegan and Operation Raleigh participants, Greg Jackson, Kevin Kennealley, Neville Marchant, Bruce Maslin, Anne Nevin, Ann Ousey, Marianne Raynal, Denis Reid, Moya Smith, Tim Willing, Ingrid Wager, Alan Winchester.

Elizabeth Adamek and Teresa Forner prepared the manuscript and arranged the references.

Photographs by the author except for the following:

Broome Botanical Society, 29, 98, 158, 179, 182, 194, 211, 227, 242, 244; Pat Dundas, 36; Bruce Gray, 80; Steven Hopper,119, 121, 122, 123, 124, 131; Arpad Kalotas, 55(b), 170; Kevin Kennealley, 41, 93, 128(a) ; King's Park and Botanic Garden, 22, 43, 44, 46, 48, 49, 52, 54, 61, 92, 111, 151, 203, 208, 215, 216, 225, 321, 234, 254, 263; Robin Maher, 41, 133, 140, 147(b), 197, 204, 240; Bruce Maslin,6, 8, 10, 13, 19, 24, 25; Eric McCrumm, 37; Philippa Nikulinsky, 252; Marcello Pennacchio, 116(a); Guy Richmond, 113, 116(b); Moya Smith, 60, 77, 103, 129, 145, 146, 148, 155, 156, 175, 178, 217, 241, 247, 264; Jan Swinney, 23, 24, 34, 38, 66, 89, 100, 125, 167, 171, 176, 177, 190, 226, 257(b), 260; Greg Wallace, 50.

INTRODUCTION

Plants were the earliest form of life on earth. One or another plant lives in most environments and on all continents. Only the polar ice caps lack plants, but even there one can find at their edges, mosses, lichens and flowering plants. Since they first appeared on earth, animals and humans too have used plants as sources of the body building foods, the calories and the vitamins that regulate metabolism. Some plants even parasitise others, using the host plants' fluids to their own ends. Australia has a number of plants that operate in this way including the Sandalwoods, the Mistletoes and the Native Cherries.

Anyone who has travelled through the bush with Aboriginal people or who has sat beside a water-hole listening to bushmen or farmers yarning could not help noticing the strong interest these people have in plants. Whether for stock fodder, traditional medicine, food to prevent "doing a perish" or as part of the standard diet, plants provide a conversation topic just as controversial as the weather. More recently, a wider group of Australians has learnt of the interest the armed services have in plants as supplementary rations for servicemen and women through a popular programme on the Australian Broadcasting Commission's television network. This presentation brought the military quest for information about bush foods into many Australian homes and gave renewed interest in plants to many individuals. A number of the anecdotes in these programmes were gleaned from Aboriginal land owners who maintained knowledge of the resources in their particular area.

My interest in plants that could be eaten, made into string, used to weave baskets or put to any purpose other than that of simple horticultural admiration began in the 1960's. With relatives and friends I roamed the banks of the Shoalhaven River in New South Wales and together we marvelled at the rock shelters painted by Aboriginal people. A handful of elderly Aboriginal men and women had divulged the location of these places to me and in doing so had recounted stories of the way of life of their ancestors. The tales were fascinating and amplified my interest in the abandoned lifeways of those who had originally occupied the rocky areas we rambled over. As I continued to explore other parts of Australia, I was always eager to discover how Aborigines and early European settlers had extracted their livelihood in each region. Working for the Department of Aboriginal Sites in the Western Australian Museum, I visited many remote areas of the state not normally visited by travellers. My field notes from this time are crammed with information about plants and animals as well as the archaeological, anthropological and ethnographic information I was gathering simultaneously.

Most of the plants and their uses mentioned here have been recorded by other workers in the field of ethnobotany. This is the first time they have all been gathered in one place and in a form that will be useful for non-botanists. An Aboriginal taxonomy of the featured plants exists in parallel to the European system. Discovering the existence of this system was no great surprise but it was fascinating to learn why various Aboriginal divisions between species of plant were not necessarily the same as those used by European botanists. Two plants considered by botanists to be different species in the same genus were classified as the same species by Aborigines, and vice versa. This field needs a volume of its own, and is not discussed here. Secondly there was a greater diversity in uses for plants than I had imagined. Take just one group, the wattles. With a distribution that fills every imaginable ecological niche, it is not surprising that a large number of wattles have been and are used by Aboriginal people. Acacias provide wood for artefacts and fires, leaves for medicine and poisons, seeds and gum for food. Edible galls occur on a number of species, and the leathery bark provides straps and twines for tying. Bushmen too have used the roasted seeds to make damper and bush coffee and the strong branches to patch broken car springs when they were in a spot of bother. Some toxic principles have been isolated from wattle seeds, and random experimentation with the seeds as food is discouraged. Sadly, few Australian plants have provided a focus for research and development as commercial crops, although this trend is now reversing. Macadamias from the eastern brush forests are one example and Quandongs from the arid interior another. Some Australian plants are being examined for their capacity to combat disease and illness following the earlier model of Pitjuri that provided a chemical model for a substance used in tablets which counter sea-sickness.

During the last fifteen years I have accepted many requests to lecture on useful plants to students undertaking "survival" courses. Some of these have been for military personnel and others for civilians. In both cases it has been my intention to provide the students with information about a number of plants that they may encounter in remote parts of Australia. It was not so much that they needed this information to prevent starvation if stranded in an isolated locality, but more to provide them with a number of plant friends and acquaintances. Hopefully these friends could provide a little psychological encouragement if one became stranded or "geographically embarrassed" as a fellow survival instructor liked to say.

Many of these students and other friends have asked for some of the information I have gathered to be compiled into a book. This publication is a response to their requests. It has only been possible though the combined efforts of a number of people whose contribution is acknowledged elsewhere. My greatest thanks must go however to the people who shared their knowledge and agreed to it being recorded for the benefit of others. I am pleased to acknowledge the contribution of Aboriginal people from numerous communities in northern and central Australia; of the station owners, stockmen and pastoral workers of the same area; of military personnel who facilitated access to remote Australia and the colleagues who have freely shared their research notes with me.

Finally a warning. Many plant products used as food by Aborigines needed a complex series of processing steps to render them edible. For example, fruits from the cycads (*Zamiaceae*) need a complex process of leaching and cooking to remove the toxins they contain before they can be safely eaten. There are plenty of other plants that require similar treatment.

Numerical references are provided to other published references to uses for the plants selected for this book. These are nowhere near exhaustive. I have tried to provide a reference which corroborates information I gathered, but sometimes there is none. Plants have been named following the usage in the various regional floras mentioned in the bibliography.

CRAB'S EYE

Scientific Name: *Abrus precatorius*

Aboriginal Name: DO-ANJIN-JIN (Broome)

References: 60, 102, 113

Habitat: Sandy soils, often just behind beaches in tropical areas.

Size and Form: Perennial woody climbing shrub, found twining around other plants.

Leaves: Compound, made up of narrow bright-green leaflets about 12 mm long. Arranged in opposite pairs.

Flowers: Pink, white or purple 'peaflower' blossoms, up to 2.5 cm wide, occurring in clusters.

Fruit: Dark-grey or brown pods, 40 – 50 mm long, 12 mm wide. Splitting open when ripe, to reveal several scarlet seeds, each with a large black spot, giving the name 'crab's eye'.

Uses: The seeds have been used for the treatment of trachoma, and to decorate wooden artefacts. The roots are used in India as a substitute for liquorice and for treating coughs and diarrhoea.

Comments: The seeds are extremely poisonous. Deaths have occurred following ingestion.

RASPBERRY JAM WATTLE

Scientific Name:	*Acacia acuminata*
Aboriginal Name:	MANGARD (south-west Western Australia)
References:	6, 12, 13, 18, 47, 86, 95
Habitat:	Very common in semi-arid woodlands.
Size and Form:	Shrub or small tree, 3 – 5 m or taller. Wide spreading crown and short cracked trunk.
Leaves:	Long and narrow, sometimes curved, 6.5 – 25 cm long, 0.25 cm wide. Many fine longitudinal veins, top margin usually fringed with short white hairs.
Flowers:	Yellow spikes, 13 – 25 mm long, 6 – 7 mm wide. Borne in pairs on the flowering stem, either on very short stalks or stalkless. Flowering July – October following rains.
Fruit:	Dark-brown linear pods, almost straight, 50 – 75 mm long, 5 – 8 mm wide. Flat or rounded. Seeds dark-brown to black, often shining, oblong to oval, 3 – 6 mm long, 2 – 4 mm wide.
Uses:	The gum is edible. Ripe seeds are ground to make flour, mixed with acacia gum and cooked as a small damper. The dense heavy timber is used to make throwing clubs and spear-throwers.
Comments:	The gum, mixed with water, is kept in jars and used like honey.

FITZROY WATTLE

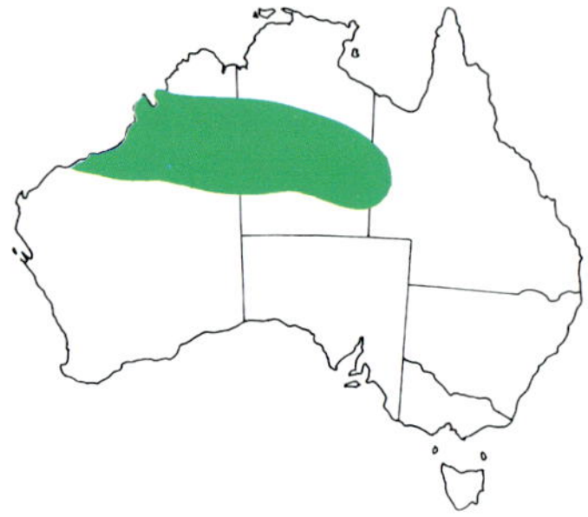

Scientific Name:	*Acacia ancistrocarpa*
Aboriginal Name:	PALPI (Port Hedland)
References:	13, 51, 60, 87
Habitat:	Sand or stony plains. Usually associated with spinifex.
Size and Form:	Spreading shrub, up to 3 m tall.
Leaves:	Linear, sometimes oval in outline, ranging from very narrow (2 mm wide) to broad (9 – 11 mm wide), 60 – 180 mm long. Bright olive-green, shiny, wrinkled when dry.
Flowers:	Clusters of 1 – 3 sparse yellowish spikes, 1 – 3 cm long.
Fruit:	Pea pod-like oblong pods, 90 mm long, 7 – 12 mm wide. Flat, curved, margins thickened, dark brown, resinous.
Uses:	The pounded phyllodes and twigs are soaked in water, and the liquid used to bathe skin sores and the forehead for relief of headaches. Leafy twigs are heated and used to soothe swellings and bruises.

BROAD-LEAF MULGA

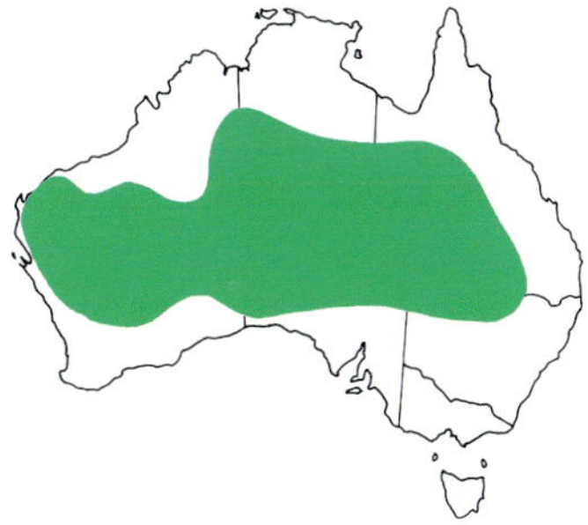

Scientific Name:	*Acacia aneura*
Aboriginal Name:	WINTAMARA (Ashburton area)
References:	13, 91, 123, 127
Habitat:	Very widespread, frequently forming pure stands on loam or sand.
Size and Form:	Shrub or tree, up to 15 m tall. Variable form.
Leaves:	Vary from needle-like (50 – 70 mm long, 2 mm wide), to flattish (50 mm long, 20 mm wide).
Flowers:	Dense bright-yellow spikes, 1 – 3 cm long, on short stems. Flowering mainly April – July, or following rains at any time of year.
Fruit:	Flattish pods, 8 cm long, 1.5 cm wide, containing several hard dark seeds.
Uses:	The seeds and gum are edible. The seeds are best ground to a paste between two stones. Following good rains, insect attack produces apricot-sized edible galls with juicy yellow flesh. The dense hard wood is used for spears and boomerangs.
Comments:	Water cavities may occasionally be found in these trees, and water is also found in the roots.

BEAD WATTLE

Scientific Name: *Acacia beauverdiana*

Aboriginal Name: PUKATI (Ashburton area)

References: 32, 60, 78, 83, 93, 126, 127

Habitat: Coarse sands and sandy gravels in semi-arid regions. Often associated with mallee eucalypts.

Size and Form: Shrub or small tree, up to 6 m tall.

Leaves: Flat linear phyllodes, 60 – 100 mm long, 2 – 4 mm wide.

Flowers: Flower heads globular, composed of golden, stalked flowers, less than 1 cm long. Flowering early spring to summer.

Fruit: Linear pods, 7 cm long, 0.3 cm wide. Constricted between seeds. Resembling a string of small beads.

Uses: The small top branches are burnt to produce a white ash, which is mixed with tobacco and chewed. Mature seeds are eaten raw, or ground to flour and cooked as cakes.

Comments: In sandhill areas, the long straight light roots of acacias are used as spear shafts.

CORK-BARK WATTLE

Scientific Name:	*Acacia coriacea*
Aboriginal Name:	ILLKILLI (Fitzroy Crossing)
References:	13, 32, 51, 114
Habitat:	Deep red sand in spinifex country.
Size and Form:	Bushy tree, up to 10 m tall. Bark thick, corky, grey and deeply furrowed.
Leaves:	Linear phyllodes, 12 – 35 cm long, 0.1 – 0.7 cm wide. Flat, covered with fine silvery hairs.
Flowers:	Golden fluffy globes on flower-bearing stalks, each composed of 30 – 40 tiny florets.
Fruit:	Legume-like pods, 15 – 23 cm long, 0.8 – 1 cm wide. Curved and twisted, leathery, becoming fibrous, covered with fine hairs. Seeds black with a red dot.
Uses:	The trunk and a lateral root are used to make ‘hooked’ boomerangs. Children thread the seeds to make necklaces. The seeds are also reported to be edible when ground with water to make porridge.
Comments:	The wood is very dense, with yellow sapwood and red heartwood.

HALL'S CREEK WATTLE

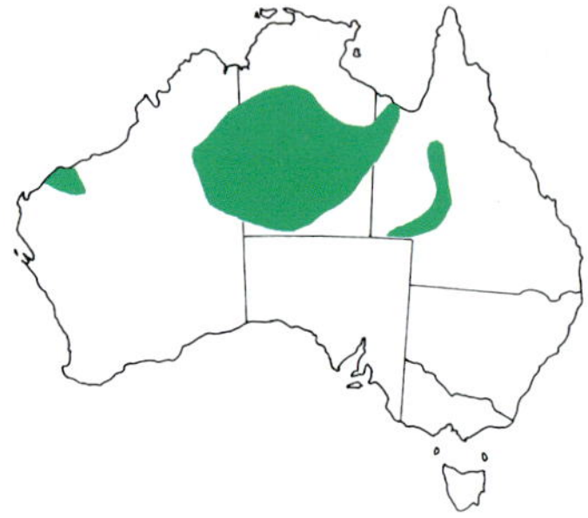

Scientific Name: *Acacia cowleana*

Aboriginal Name: PADA (Wangkatjanka Community)

References: 32, 51, 73, 87

Habitat: Usually favours sandy or stony soils.

Size and Form: Rather spindly shrub or small tree, up to 5 m tall.

Leaves: Leaf-like symmetrical phyllodes, almost sickle-shaped, 8 – 20 cm long, 0.9 – 2.4 cm wide. Young leaves often covered in soft hairs.

Flowers: Flower spikes dense, yellow, 15 – 40 mm long, on stalks 4 – 12 mm long.

Fruit: Legume-like pods, 75 mm long, 4 mm wide, curved. Firm, papery, raised over seeds but barely constricted between them.

Uses: The seeds are ground to a paste with water and eaten raw. The wood is used to make artefacts.

Comments: Damper made from the seeds is reputed to cause headaches.

HOP MULGA

Scientific Name: *Acacia craspedocarpa*

Aboriginal Name: THADANGU (Murchison area)

References: 32, 74

Habitat: Floodplains and deep soils near watercourses. Frequently occurs in thickets.

Size and Form: Shrub or small many-branched tree, up to 3 m tall.

Leaves: Usually oval, 10 mm long, 5 mm wide.

Flowers: Rod-like, up to 2 cm long, occurring singly.

Fruit: Broad flattish heavily veined pods, 10 cm long, 3 cm wide. Containing four or five flattened seeds.

Uses: The young seeds can be eaten raw.

Comments: The seeds are much sought-after by Aborigines as a 'snack' food.

CUTHBERTSON'S WATTLE

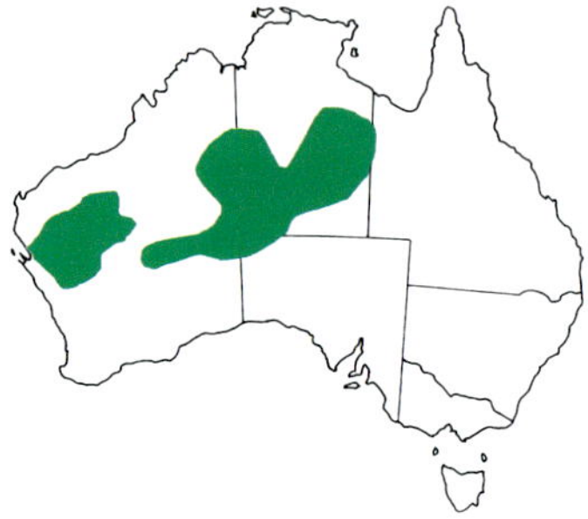

Scientific Name:	*Acacia cuthbertsonii*
Aboriginal Name:	PIRLIYI (Christmas Creek area)
References:	12, 13, 51, 60, 62, 91, 97
Habitat:	Often rocky ground.
Size and Form:	Silver-grey spreading shrub, up to 3 m tall.
Leaves:	Phyllodes narrow, oblong or tapering towards base, 4 – 8.5 cm long, up to 1 cm wide.
Flowers:	Flower spikes yellow, heads globular, 15 – 30 mm long, on stalks 5 – 10 mm long.
Fruit:	Legume-like oblong pods, 16 cm long, 1 – 2 cm wide. Normally curved, woody, yellowish, hairless. Coarsely wrinkled when dry.
Uses:	A bark decoction is used for the relief of toothache and rheumatism. Strips of bark are used as bandages and as string for tying and decoration.

STICKY WATTLE

Scientific Name:	*Acacia dictyophleba*
Aboriginal Name:	PURRPUN (Christmas Creek area)
References:	74, 76, 114, 127
Habitat:	Open woodland or hummocky grassland, mainly in red siliceous sands but also in rocky ground.
Size and Form:	Bushy shrub, 1 – 4 m tall, with several stems spreading from base.
Leaves:	Phyllodes tapered towards base, 2 – 7 cm long, 0.5 – 1.5 cm wide. Straight or slightly curved, leathery, sticky and smooth.
Flowers:	Globular yellow flower-heads, composed of 40 – 60 florets. Arranged singly or in pairs. Flowering April – September.
Fruit:	Legume-like oblong pods, 5 – 9 cm long, 1.2 – 1.5 cm wide. Flattish but raised over seeds. Thin, light-brown, often with a shining varnished appearance.
Uses:	The leaves are boiled (or soaked) in water until the solution looks like weak tea. The resulting liquid is drunk to alleviate colds, coughs and headaches. The liquid can probably also serve as a 'wash' (see *Acacia ancistrocarpa*).

IRONWOOD

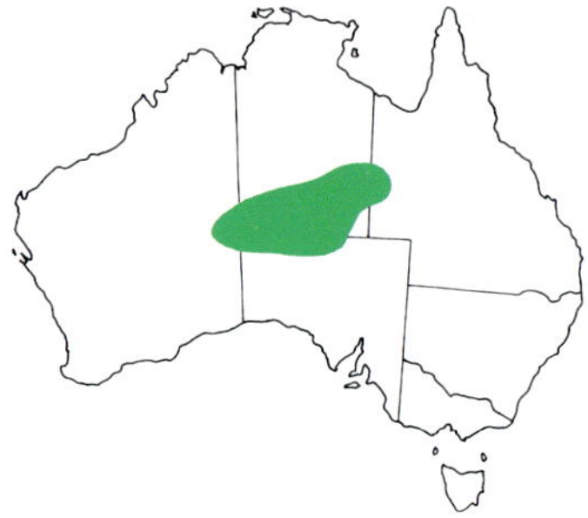

Scientific Name:	*Acacia estrophiolata*
Aboriginal Name:	TJAAWU (Alice Springs)
References:	12, 74, 97
Habitat:	Favours open plains on well-drained sandy soils.
Size and Form:	Tree, 4 – 15 m tall, with erect branches supporting long thin pendulous foliage.
Leaves:	Linear flat slightly curved phyllodes, ending in a short blunt point, 4 – 10 cm long, 0.2 – 0.3 cm wide. Smooth, flexible, three-veined.
Flowers:	Yellow spherical flower-heads, attached to short, unbranched stalks. Occurring singly or in pairs amongst the foliage.
Fruit:	Flattened linear pods, 2 – 10 cm long. Constricted between seeds. Seeds attached longitudinally to pod by a thin thread-like stalk.
Uses:	The gum is applied to skin lesions as a dressing. Bark from the small roots is crushed and soaked in hot water to make a dark-red solution, which is applied to sores and burns. Bark strips are used as bandages.
Comments:	It has been suggested that baler shells (*Melo* species), other large marine shells and even wooden dishes might have been used as containers in which to heat water prior to the acquisition of 'billies', but their effectiveness has not been tested.

RED CURLY-BARK

Scientific Name: *Acacia grasbyi*

Aboriginal Name: MINNIRITCHI (Meekatharra)

References: 13, 36, 47, 51, 76, 127

Habitat: Rocky areas in 'red sand' country.

Size and Form: Shrub or tree, up to 6 m tall. Red curly bark, peeling in strips from the trunk.

Leaves: Phyllodes 4 – 9 mm long, 1 mm wide. Usually round in cross-section, fairly stiff and slightly curved.

Flowers: Dense yellow spikes, up to 3 cm long.

Fruit: Reddish pods, straight or curved, 10 cm long, 0.5 – 0.8 cm wide. Seeds pale brown.

Uses: The wood is used to make spear points for spears with shafts of acacia root. It is also used for clubs and boomerangs.

Comments: The wood is strong, dense and attractively coloured, having a dark-red centre and cream outer layer. The desert snakewood (*Acacia eremaea*; BILYA, Murchison area) and red mulga (*Acacia cyperophylla*) are also used similarly, along with many other acacia species of the arid lands.

CANDELABRA WATTLE

Scientific Name:	*Acacia holosericea*
Aboriginal Name:	KURKANYAN (Pilbara)
References:	43, 47, 62, 87
Habitat:	In dense clumps, in disturbed areas or near watercourses.
Size and Form:	Straggly spreading shrub or tree, up to 8 m tall.
Leaves:	Grey-green, 10 cm long, 1 – 2 cm wide, with two or three prominent veins.
Flowers:	Yellow spikes, 3 – 6 cm long.
Fruit:	Pods borne in tight clusters, twisted and curled when mature. Seeds black, surrounded by a yellow or red attachment to the pod.
Uses:	The small seeds are edible.
Comments:	In wetter locations, the leaves are larger and darker green than in dry locations, where they take on a silvery appearance.

CAMEL WATTLE

Scientific Name:	*Acacia inaequilatera*
Aboriginal Name:	PADERI (Port Hedland)
References:	6, 47, 51
Habitat:	Common on hills and plains in sandy, gravelly soils.
Size and Form:	Tree, up to 5 m tall. Corky bark, often burnt black.
Leaves:	Grey-green, stout with offset central vein and spiky margins, 4 cm long.
Flowers:	Globular flower heads. Petals purple.
Fruit:	Pods curved, with a papery or leathery feel, 8 cm long, 1.5 cm wide.
Uses:	The red inner bark is pounded, soaked or boiled until the liquid becomes red and thick, then used to bathe skin sores, cuts and rashes. It has an astringent, drying action.
Comments:	Often confused with shrubby forms of the ranji wattle (*Acacia pyrifolia*).

WITCHETTY BUSH

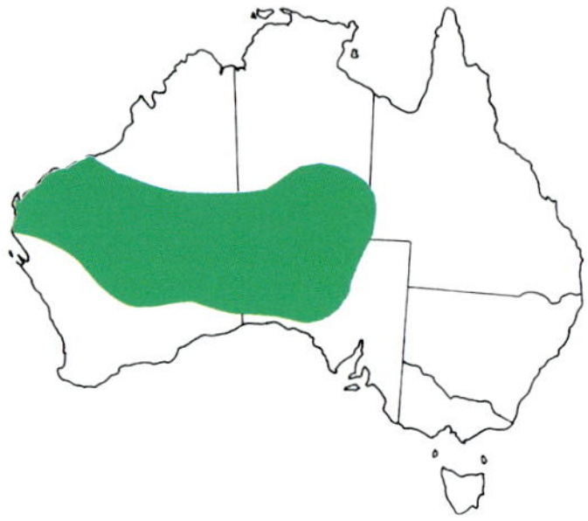

Scientific Name:	*Acacia kempeana*
Aboriginal Name:	WINTALKA (central desert)
References:	16, 51, 114
Habitat:	Red siliceous sand and shallow calcareous loam, or rocky sandstone.
Size and Form:	Shrub, 2 – 4 m tall. Smooth, spreading, dull grey-green.
Leaves:	Oblong phyllodes, tapering towards tip, 3 – 7 cm long, 0.4 – 1.2 cm wide. Slightly curved, flat, leathery, sticky when young.
Flowers:	Dense mid-yellow spikes, one or two per stem, 1 – 2 cm long. Stalks smooth, 0.5 – 1.2 cm long.
Fruit:	Oblong flat brown pods, 3 – 5 cm long, 1 – 1.5 cm wide. Papery, smooth or sticky.
Uses:	The leaves are chewed or used to make a decoction to treat bad colds. They are also boiled in water and used as a wash for wounds.
Comments:	The surface roots harbour large insect larvae known as witchetty grubs, which are considered a great delicacy. They are extracted from their tunnels with small hooked twigs and baked in ashes or eaten raw.

UMBRELLA BUSH

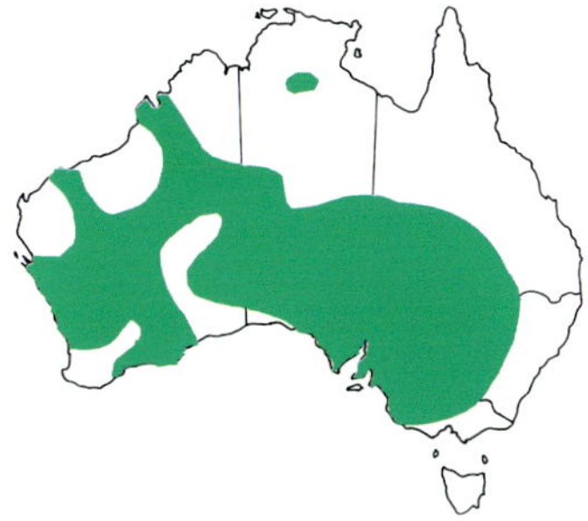

Scientific Name:	*Acacia ligulata*
Aboriginal Name:	WATARRKA (central desert)
References:	60, 78, 83, 97, 123, 127
Habitat:	Widespread on dry alkaline soils or coastal dunes.
Size and Form:	Shrub, up to 3 m tall.
Leaves:	Narrow linear or oblong leaf-like phyllodes, with a small hard point. Tapered at both ends, thick, up to 10 cm long, 2 cm wide.
Flowers:	Profusion of yellow-orange balls, 5 – 6 mm in diameter. Borne singly or in groups of 2 – 5 in leaf axils. Flowering late winter through spring.
Fruit:	Hard pods, 5 – 10 cm long, constricted and brittle between individual seeds.
Uses:	The seeds are ground and eaten. A bark decoction is used as a general medicine for dizziness, nerves and fits. Branches are used to 'smoke' people for general sickness and for women following childbirth.
Comments:	There may be several forms of this wattle, or a group of similar species not yet botanically differentiated.

BOWGADA

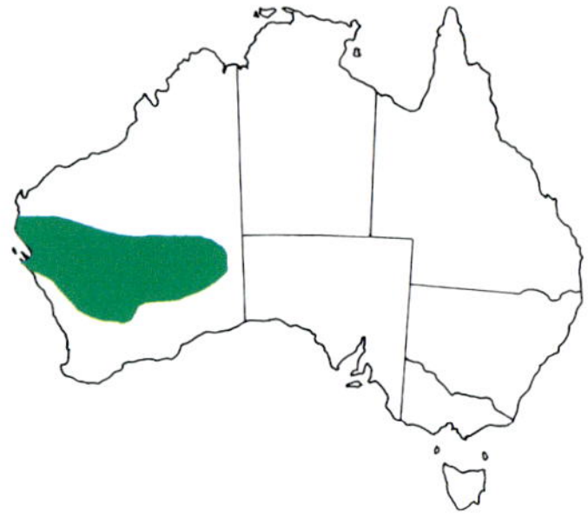

Scientific Name:	*Acacia linophylla*
Aboriginal Name:	BOWGADA (Murchison area)
References:	5, 13, 16, 62, 63, 74, 76, 78, 127
Habitat:	Open scrub, tall shrubland or low woodland.
Size and Form:	Shrub, 2 – 6 m tall, often 2 – 6 m or more wide. Branching into several slender stems.
Leaves:	Linear, flat, 8 – 20 cm long, 0.15 – 0.3 cm wide. Tipped with a short point and covered with fine hairs.
Flowers:	Dense yellow spikes, 1 – 2 cm long, in bunches of one or two. Each flower composed of five florets.
Fruit:	Cylindrical pods, 7 – 13 cm long, 0.5 – 0.8 cm wide. Woody when mature. Seeds arranged longitudinally in pod.
Uses:	The pods are steamed and eaten when green and young.
Comments:	Also known as WANDARRIE, and gives this name to a vegetation regime in the Murchison area. Cannot satisfactorily be distinguished from *Acacia ramulosa*.

TURPENTINE WATTLE

Scientific Name:	*Acacia lysiphloia*
Aboriginal Name:	WATARUDU (Balgo area)
References:	12, 87, 91, 102
Habitat:	Common on red sand in spinifex communities.
Size and Form:	Shrub, up to 5 m tall. Bark reddish and loose, peeling upwards in curls.
Leaves:	Phyllodes narrow, pointed, 1 – 4 cm long.
Flowers:	Bright orange-yellow spikes, 3 cm long.
Fruit:	Pods 4 – 6 cm long, 0.6 cm wide. Sticky with a scented resin. Curled when mature.
Uses:	Leafy green branches used for 'smoking' to treat general sickness.
Comments:	The name MINIRITCHIE is loosely applied to wattles having bark that curls away from the branches in longitudinal rolls, but probably refers to an Aboriginal reference to the spike-like leaves of some species.

MANNA WATTLE

Scientific Name: *Acacia microbotrya*

Aboriginal Name: KALYANG (south-west Western Australia)

References: 12, 18, 34, 72, 78, 95, 126, 127

Habitat: Widespread in sandy loams and in spinifex country. Fast-growing in warmer areas.

Size and Form: Bushy shrub or small tree, 2 – 6 m tall, up to 5 m wide. Bark rough and grey at base.

Leaves: Green-grey pendulous curved lance-shaped phyllodes. Extremely variable in shape and size, mostly 7 – 13 cm long, 0.6 – 1.5 cm wide.

Flowers: Small bright-yellow sprays, 4 – 6 mm wide, each composed of about 20 flowers on stalks 3 – 4 mm long.

Fruit: Dark-brown finely veined pods, 10 – 20 cm long, 0.5 – 0.7 cm wide. Constricted between seeds, slightly thickened margins. Seeds black, oblong, flat, 5 – 7 mm long, 3 – 4 mm wide.

Uses: The seeds are pounded to make a flour, which is cooked as dampers or unleavened bread. The tree produces great quantities of edible gum in dry times.

HILL WATTLE

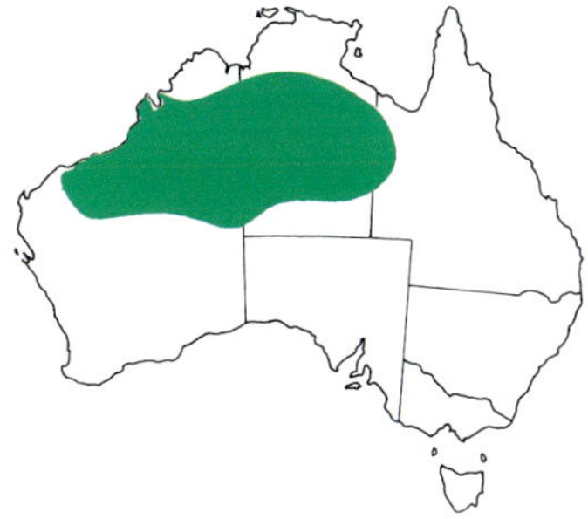

Scientific Name: *Acacia monticola*

Aboriginal Name: GAWAN (Broome)

References: 12, 51, 60, 91, 95, 102

Habitat: Favours shallow sandy or stony soils, and ironstone on rocky ranges.

Size and Form: Shrub, up to 4 m tall. Dry reddish bark curling off in thin strips.

Leaves: Leaf-like phyllodes, 4 cm long, 2 cm wide. Sticky, some short hairs on margins, semi-transparent.

Flowers: Large loose yellow balls (or rarely spikes), 9 mm in diameter, each composed of 15 – 20 flowers.

Fruit: Flat sticky red-brown pods. Seeds brown, shining, 5 – 6 mm long, 3.5 – 4.5 mm wide, 2 mm thick.

Uses: Used to treat coughs and colds. Mashed roots are soaked or boiled and the resulting solution drunk or used to bathe the body.

Comments: A decorative shrub, but seldom used in horticulture.

MURRAY'S WATTLE

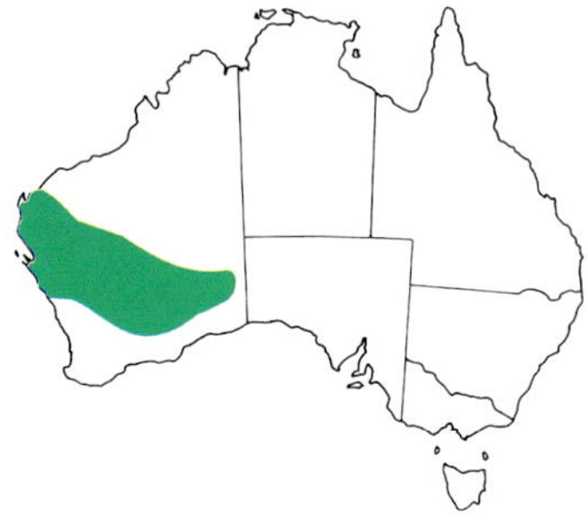

Scientific Name:	*Acacia murrayana*
Aboriginal Name:	TJUNTALA (western desert)
References:	13, 16, 74, 78, 93, 95, 123, 127
Habitat:	Widespread in arid scrubland and sandhill country in sunny positions.
Size and Form:	Bushy shrub or small tree, 2 – 5 m tall. Spreading or erect, smooth-barked, spreading crown.
Leaves:	Variable, pale green to blue-green, flat, thick with thickened yellowish margins. Usually 5 – 18 cm long, 0.2 – 0.7 cm wide.
Flowers:	Bright yellow balls, 7 – 9 mm in diameter. On slender stalks, forming open-branched bunches.
Fruit:	Light-brown to red papery pods, up to 7 cm long, 1 cm wide. Raised over seeds and constricted between them. Seeds black.
Uses:	The seeds are softened by soaking in water, ground to a paste while still damp, scooped from the grinding stone with cupped fingers and eaten raw.
Comments:	Edible insect larvae, 'bardie grubs' or MAKU (Kalgoorlie), are found in the roots and branches. Piles of 'sawdust' reveal their presence, and they are pulled from their tunnels with small hooked twigs.

NEEDLEWOOD

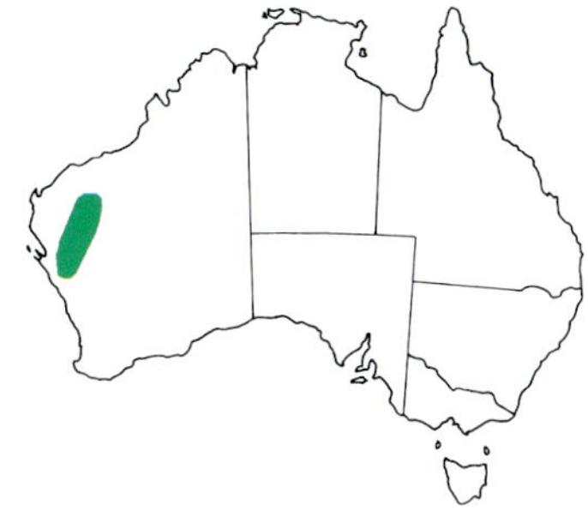

Scientific Name:	*Acacia palustris*
Aboriginal Name:	GOWA (Murchison area)
References:	76, 127
Habitat:	Open country scattered with snakewood and other acacias.
Size and Form:	Tree, 2 – 3 m tall. Bark black.
Leaves:	'Needles' to about 10 cm long, deflexed downwards.
Flowers:	Cylindrical, yellow, 1 cm long, on short stems.
Fruit:	Flat, slightly curved pods, 8 – 20 cm long, constricted between seeds.
Uses:	The young pods are cooked and eaten whole. Mature seeds are ground to a flour and cooked as damper.

SOAP WATTLE

Scientific Name: *Acacia pellita*

Aboriginal Name: KANKULANG (Kununurra)

References: 58, 65

Habitat: A wide variety of well-drained soils on plains.

Size and Form: Tree, up to 5 m tall but usually smaller. Bark smooth, dark brown.

Leaves: Broad, 15 cm long, 6 cm wide. Grey-green, furry, appear stiffly held. Soft and velvety to the touch.

Flowers: Cylindrical golden spikes, up to 70 mm long, 7 mm wide. One or two per leaf axil.

Fruit: Narrow strap-like pods, 5 mm wide. Curled in a tight spiral, containing black seeds.

Uses: The leaves and seeds are soaked and the liquid used as a wash to treat sores and relieve pains. Smoke of burning leaves is used to quieten children. The branches can be rapidly drawn through pools to stupefy fish.

Comments: The green pods produce a soapy lather when rubbed vigorously between wet hands.

GOLDEN WREATH WATTLE

Scientific Name:	*Acacia saligna*
Aboriginal Name:	WUANGA (south-west Western Australia)
References:	12, 72, 93, 114
Habitat:	Widespread and common, favouring calcareous sands or leached, siliceous soils in coastal areas.
Size and Form:	Dense bushy shrub or small tree, 3 – 8 m tall. Bark smooth, red-brown.
Leaves:	Phyllodes linear, tapering gradually towards tip, 8 – 25 cm long, 0.2 – 5 cm wide, either straight or sickle-shaped.
Flowers:	Flowering stem composed of a main axis carrying five flower-heads on lateral stalks. Flower heads globular, deep-yellow to orange, composed of 25 – 75 florets.
Fruit:	Pea pod-like linear pods, 5 – 14 cm long, 0.5 – 0.6 cm wide. Straight or slightly curved, smooth, brown. Seeds arranged longitudinally in pod.
Uses:	The seeds are edible.
Comments:	It is probably these seeds that were crushed to flour and eaten with pounded root bark from various eucalypts, including *Eucalyptus cornuta*.

DEAD FINISH

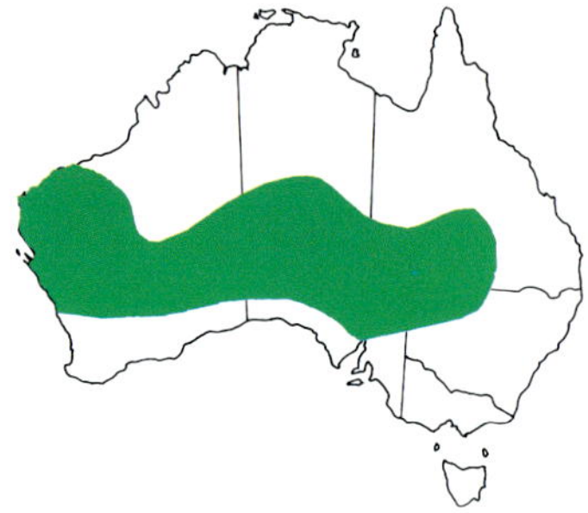

Scientific Name: *Acacia tetragonophylla*

Aboriginal Name: KURARA (Murchison and Goldfields area)

References: 51, 63, 82, 93

Habitat: Favours sandy soils near watercourses, but found in a variety of habitats.

Size and Form: Diffuse shrub or tree, up to 5 m tall. Branches normally smooth.

Leaves: Phyllodes needle-like, tapering to a stiff sharp point, 10 – 60 mm long, 1 mm wide, with four longitudinal channels. Occurring in clusters.

Flowers: Flower heads yellow, globular, dense, composed of about 50 flowers. Stalks 1 – 3 cm long.

Fruit: Pods dark-brown, 13 cm long, 0.4 – 0.6 cm wide. Normally curved or twisted. Seeds attached to pods by yellow stalks, which encircle them.

Uses: The seeds are eaten raw or, when mature, are pounded and eaten uncooked. They have a pleasant nutty flavour. The dense wood is favoured for clubs.

Comments: The points of needle-like phyllodes of this and similar species are pushed beneath warts, which then wither away.

POVERTY BUSH

Scientific Name: *Acacia translucens*

Aboriginal Name: BANMUNG (Port Hedland)

References: 13, 60, 63, 87, 95

Habitat: Prefers deep red sands overlying sandstone ridges.

Size and Form: Semi-prostrate spreading shrub, up to 1 m tall.

Leaves: Phyllodes almost oval to club-shaped, 1 – 2 cm long. Slightly resinous, leathery texture.

Flowers: Yellow globose flower-heads, 1 cm in diameter, each composed of 30 – 40 florets. Each flower-head on a stout, hairless, ribbed stalk, 1.2 – 2 cm long.

Fruit: Flat thick woody pods, 4 – 6 cm long. Seeds greyish, oblong, arranged obliquely in pod.

Uses: The leaves and twigs are mashed in water and the liquid used to bathe skin sores and as a 'wash' for treating headaches.

Comments: In the Pilbara, where this plant is present in large numbers in grazing areas, it indicates over-grazing and poor range conditions. Hence the locally used common name, which is also applied to *Eremophila* spp. in the Murchison.

PINDAN WATTLE

Scientific Name: *Acacia tumida*

Aboriginal Name: NALDUPA (Wangkatjunka Community)

References: 47, 51, 87, 102

Habitat: Common in tropical regions in a variety of habitats.

Size and Form: Shrub or tree, 7 – 10 m tall. Bark very variable, often silvery on young branches, turning reddish with age.

Leaves: Phyllodes sickle-shaped, 23 cm long, 5 cm wide. Veins arising at the phyllode base and running parallel to the margins.

Flowers: Yellow cylindrical spikes, 70 mm long, 7 mm wide.

Fruit: Mature pods fragile, wrinkled or twisted, 6 – 11 cm long.

Uses: The young pods are steamed and eaten raw.

Comments: Some Aboriginal groups claimed that the mature seeds of this species were not crushed for damper, but other groups did use it in this way. Some toxic principles have been isolated from wattle seeds, and experimentation with the seeds as food is discouraged.

GUNDABLUEY WATTLE

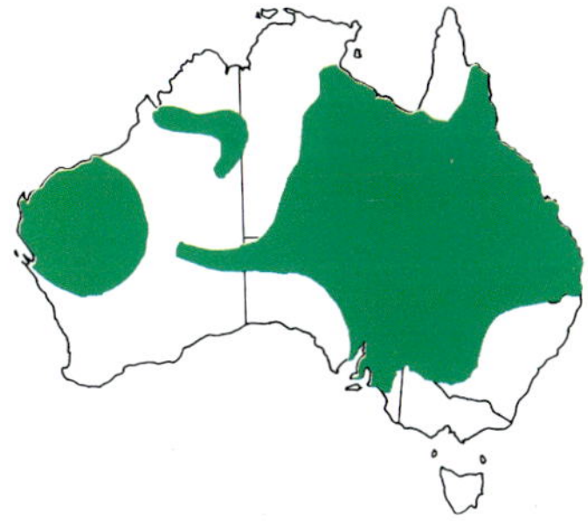

Scientific Name: *Acacia victoriae*

Aboriginal Name: WALIPUTA (Murchison area)

References: 12, 43, 62, 63, 74, 82, 87, 95, 114, 123, 127

Habitat: Widespread in open shrubby woodlands or in dense thickets on scrubland.

Size and Form: Bushy shrub or small tree, up to 5 m tall. Usually spiny and light green.

Leaves: Phyllodes linear, flat, with single vein, up to 5 cm long, 1 cm wide. Pale green when juvenile, changing to waxy grey-green. Borne singly at different heights on stem, with one or two spines, 0.8 cm long.

Flowers: Globular cream to pale-yellow flower-heads, 8 mm in diameter. On branched stalks.

Fruit: Short curved green-brown pods, 2 – 3 cm long, 1 – 1.3 cm wide. Seeds oval to circular, attached transversely to pod by short stalks.

Uses: The seeds are used to make ornamental necklaces, holes being burnt through them with wire. Edible grubs are found in the roots and stems. The seeds can be used as an emergency food after parching and grinding to a paste with water and cooking as a damper.

SNAKEWOOD

Scientific Name: *Acacia xiphophylla*

Aboriginal Name: MARRUWA (Roebourne area)

References: 13, 76, 127

Habitat: Favours red soils, growing in groves surrounding creek beds on plains.

Size and Form: Low stunted bent tree, up to 4 m tall. Outer bark of trunk fibrous and deeply fissured into 'twists'.

Leaves: Blue-green, flat, 2 cm long, with prominent veins.

Flowers: Pale yellow, rod-like, 1.5 cm long, solitary.

Fruit: Pods up to 11 cm long, 0.6 cm wide. More or less straight. Smooth and light brown.

Uses: Curved limbs are used for making boomerangs. The gum and seeds are edible. In sandy country, the long straight roots are dug from just beneath the surface, for use as spear shafts.

Comments: The gum is collected, rolled into a ball, covered with paperbark from melaleuca and tied with bush twine for transport and storage.

BOAB

Scientific Name:	*Adansonia gregorii*
Aboriginal Name:	DJUNGURI (West Kimberley)
References:	24, 87
Habitat:	Sandy soils in low-rainfall areas, mainly beside drainage lines and creeks.
Size and Form:	Tree, up to 16 m tall. Remains leafless for much of the dry season. Smooth yellowish green to brown bark. Trunks of young trees slightly swollen towards base. Older specimens have almost globular or bottle-shaped trunks, up to 3 m in diameter. Bare of branches for first 3 – 4 m above ground level.
Leaves:	Compound, consisting of 5 – 7 leaflets. Each leaf elliptic, tapering towards base, 13 cm long, 6 cm wide.
Flowers:	Large, up to 12 cm in diameter. Five oblong cream-white petals, almost covered by protruding stamens. Often blooming with onset of rains.
Fruit:	Velvety ovoid 'nut', 10 – 20 cm long, containing numerous kidney-shaped seeds embedded in white pith.
Uses:	The fruits, including seeds, are eaten raw or roasted. The leaves and pithy parts of young branches are eaten as a vegetable, and a drink is made from the sap. Strong twine is made from the fibrous bark of young trees.
Comments:	The only Australian representative of this African-Australian tree group. Although the fibrous wood contains much water, this cannot be 'tapped' as free-flowing liquid. Lengths of the root yield water if drained over a container.

KAPOK BUSH

Scientific Name:	*Aerva javanica*
Aboriginal Name:	PILARI (Pilbara)
References:	12, 51
Habitat:	Widespread in temperate and tropical sub-desert areas on a wide range of soils.
Size and Form:	Many-branched perennial herb, up to 1 m tall, covered in short soft hairs.
Leaves:	Ovate, tapering towards the tip, 2 – 7 cm long. Borne singly at different heights on the axis, or sometimes in clusters.
Flowers:	White, dense, arranged in panicles. Sepals white and densely woolly.
Fruit:	Seeds tiny, smooth, brown, lenticular and surrounded with fluff.
Uses:	The fluff surrounding the seeds is collected in large quantities as stuffing for pillows. It can probably be used as tinder in fire-lighting.
Comments:	An introduced species. Aborigines may have followed Europeans in their use of this plant.

SNAKESKIN LILY

Scientific Name: *Amorphophallus galbra*

Aboriginal Name: (Not collected)

References: 2, 25, 32, 113

Habitat: Vine thickets and savanna areas, often in loamy soils.

Size and Form: Herbaceous perennial, 0.5 – 1 m tall. Root tuber roughly ovoid, up to 10 cm wide.

Leaves: Solitary, 60 cm wide. Divided into three main sections about 30 cm wide. These in turn are divided into five or more leaflets. Stalk spotted.

Flowers: Stem up to 60 cm tall, mottled. Spadix yellow, 12 cm long, flowers occurring on the basal 6 cm.

Fruit: Red berries, 1 cm in diameter.

Uses: The tubers, leaf stalks and fruit are edible, after being thoroughly baked.

Comments: Dries back to the underground tuber during the dry season.

WILD GRAPE

Scientific Name: *Ampelocissus acetosa*

Aboriginal Name: KANERA (Derby)

References: 12, 32, 54, 65

Habitat: Open forest, edges of monsoon vine thickets and sand dunes near swamps.

Size and Form: Vine, climbing by means of tendrils. Stem pale, almost white, covered with short dark prickles.

Leaves: Bright green. Divided into seven leaflets: two sets of three, plus one main leaflet.

Flowers: Small, deep maroon, petals absent. Five yellow stamens protruding at right angles to the ovary. Borne in clusters.

Fruit: Grape-like clusters, currant-sized, black when ripe. (January – March.)

Uses: The fruit is edible raw. The large fibrous storage organ, found just beneath the soil, is edible but bitter.

Comments: The plant dies back to dormant rootstock during summer. Flowers may appear before the new leaves are fully formed.

SOUTHERN MISTLETOE

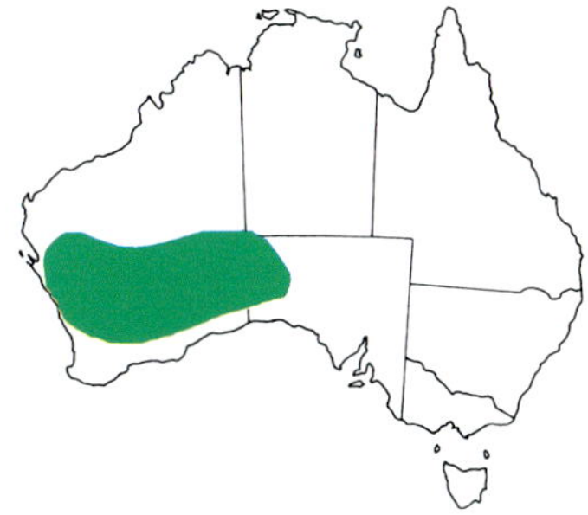

Scientific Name:	*Amyema fitzgeraldii*
Aboriginal Name:	(Not collected)
References:	51, 72
Habitat:	Widespread as a parasite of the stems and branches of various tree and shrub species, particularly *Acacia acuminata*.
Size and Form:	Erect or spreading mass, about 50 cm in diameter.
Leaves:	Oblong to oval, 2 – 5 cm long, glaucous.
Flowers:	Stalk 5 – 15 mm long, bearing 3 – 5 tubular greenish stalkless flowers. Petals slender, 1 mm wide, 18 – 26 mm long, terminating in a point.
Fruit:	Globular, succulent, pink, 6 mm in diameter.
Uses:	The berries are edible.
Comments:	European mistletoes bear poisonous berries, but there are no reported cases of poisoning from Australian species.

A MISTLETOE

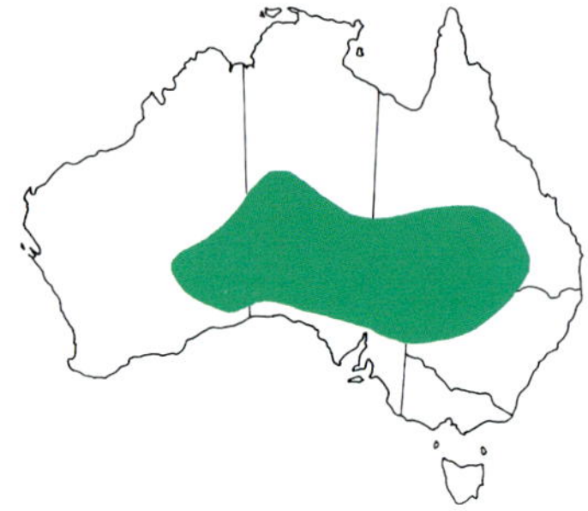

Scientific Name: *Amyema maidenii subsp. maidenii*

Aboriginal Name: NYINKIRNI (central Australia)

References: 17, 123

Habitat: Widespread as a parasite of the stems and branches of many tree and shrub species.

Size and Form: Densely clumped semi-erect or slightly pendulous mass of thin branchlets, up to 1 m in diameter.

Leaves: Oval, tapering at base, 2 – 6 cm long. Distinctive grey-green.

Flowers: Petals in mature bud slender, terminating in distinct point, 18 – 26 mm long, dark green. Stalk 10 – 30 mm long.

Fruit: Nearly globular, 6 mm long, grey-green outer skin. Sticky or mucilaginous when skin broken.

Uses: The fruit is edible and sweet. Prepared by popping it out of the outer grey-green skin.

Comments: Not an important food source, but the sweet, pleasant fruit is eagerly sought by children.

BLACKCURRANT

Scientific Name: *Antidesma ghaesembilla*

Aboriginal Name: YANGU (Kalumburu)

References: 2, 24, 32, 65

Habitat: Confined to coastal scrubs.

Size and Form: Small straggly tree, up to 3 m tall. Bark grey and fissured.

Leaves: Dark green, shiny, oval, 11 cm long, 6 cm wide. Margins tending to curl upwards. Smooth, but dimpled between veins.

Flowers: Small yellow spikes, more or less globular, with many protruding stamens.

Fruit: Greenish, turning black when ripe, 5 mm long, 3 mm wide.

Uses: The fruit is edible raw, and is mainly eaten by children. Often crushed in water to make a sweet drink. Early settlers used the fruit to make jam. The fruit produces a purple dye, which is used in basketry.

Comments: Occurs widely across tropical Australia, and is eaten wherever it occurs.

SEA CELERY

Scientific Name:	*Apium prostratum*
Aboriginal Name:	(Not collected)
References:	25, 32, 63, 67
Habitat:	Common in saline coastal areas, and rarely in inland regions. Favours open sunny locations.
Size and Form:	Spreading perennial herb, up to 0.5 m tall. Stems angular, trailing along the ground or growing erect after a semi-horizontal beginning.
Leaves:	Divided into 3 – 5 leaflets. Very variable, from broad to narrow.
Flowers:	White-cream, tiny. On individual stalks of about equal length, arising in a cluster at the top of the flowering stem.
Fruit:	Globular, 2 mm in diameter, with corky ribs.
Uses:	Substitute for domestic celery. Eaten by Captain Cook's group at Endeavour River, Queensland. Also occurring in New Zealand.

YELLOW PONDLILY

Scientific Name: *Aponogeton elongatus*

Aboriginal Name: YANGUN (Kalumburu)

References: 4, 24, 63, 113

Habitat: Favours permanent freshwater pools and streams with muddy or silty bottoms.

Size and Form: Aquatic plant, mainly submerged, to about 1 m in diameter. Leaves arise from top of tuber.

Leaves: Varied, mostly submerged beneath water. Those floating have stalk 1 – 50 cm long, oval blade 10 – 15 cm long and 1.5 – 4 cm wide. Prominent veins and wavy margins.

Flowers: Composed of two yellow 'petals' enclosed within an outer sheath 15 mm long. May remain submerged. Scented.

Uses: The tubers are edible. They are eaten raw or cooked in ashes and peeled. Available during the wet season.

Comments: This plant is used in the aquarium trade.

MOSS-LEAVED HEATH

Scientific Name:	*Astroloma ciliatum*
Aboriginal Name:	KORDIN (Ravensthorpe)
References:	10, 12, 32, 68, 86
Habitat:	Light to heavy soils in various habitats.
Size and Form:	Prostrate or mat-like shrub, 30 cm tall. Branchlets covered in minute hairs.
Leaves:	Narrow, tapering to a sharp point, 10 mm long, 1 mm wide. Flat or convex, hairy, crowded along stems.
Flowers:	Tubular, 2 – 3 cm long, erect, protruding through leaves, red or deep purple-red. Flowering May – October.
Fruit:	Globular, succulent, single-seeded, 5 mm in diameter.
Uses:	The fruit is edible raw.

KICK BUSH

Scientific Name: *Astroloma pallidum*

Aboriginal Name: KADGEEGURUP (Perth)

References: 12, 32, 71

Habitat: Favours sandy or gravelly well-drained soils.

Size and Form: Prostrate or dwarf many-branched shrub, up to 25 cm tall.

Leaves: Narrow, tapering to stiff prickly point, 1.5 cm long. Hairy, toothed margins.

Flowers: Tubular, 1.5 cm long. Densely bearded petals, white or pale pink lobes.

Fruit: Globular, 0.5 cm in diameter, succulent, green.

Uses: The fruit is edible raw.

Comments: Many of the *Astroloma* spp. and related genera like Leucopogon produce edible fruits, often called 'cranberries'. These have been collected by spreading skin cloaks beneath the bushes and beating the bushes lightly with a stick.

A CRANBERRY

Scientific Name:	*Astroloma serratifolium*
Aboriginal Name:	KONDRUNG (south-west Western Australia)
References:	32, 63, 72
Habitat:	Usually in forest areas with well-drained light to medium soils.
Size and Form:	Erect dwarf shrub, 0.3 – 1 m tall. Many branches, densely leaved at ends of branchlets.
Leaves:	Oval, tapering to a stiff sharp point, up to 1 cm long, 0.5 cm wide. Margins serrated and curved downwards. Blue-green.
Flowers:	Tubular, tapering at both ends, up to 1.5 cm long. Red-pink. Produced on older wood.
Fruit:	Small, green, fleshy, with a single seed.
Uses:	The fruits are edible when soft. (They do not usually change colour when ripe.)
Comments:	Suitable for cultivation as a garden or container ornamental plant.

OLD MAN SALTBUSH

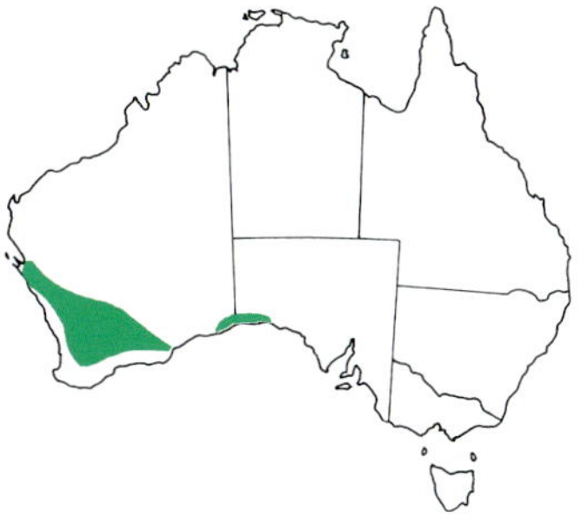

Scientific Name: *Atriplex nummularia*

Aboriginal Name: PINING (Esperance)

References: 13, 72, 74, 78, 83

Habitat: Arid low-lying areas, on clay soils.

Size and Form: Shrub, up to 3 m tall. Predominantly dioecious.

Leaves: Oval or nearly circular, 2 cm long. Blade thick and fleshy, margins smooth or with shallow teeth. On short stalks.

Flowers: Male and female flowers on different parts of plant. Insignificant. Borne in panicles at the ends of branches.

Fruit: Fruiting body stalkless, circular or somewhat fan-shaped, 5 – 15 mm long, 5 – 10 mm wide. Upper margins smooth or toothed.

Uses: The leaves can be steamed or boiled in two changes of water and used as a pot-herb. A decoction of the leaves is used to bathe skin sores, burns and wounds.

BLADDER SALTBUSH

Scientific Name: *Atriplex vesicaria*

Aboriginal Name: (Not collected)

References: 13, 51, 72, 74, 78, 83

Habitat: Widely distributed throughout drier inland Australia, on alluvial flats and gravel plains.

Size and Form: Erect shrub, up to 1 m tall.

Leaves: Blade oval, narrowing towards base, 5 – 25 mm long, 3 – 15 mm wide. Stalked. Grey-green with scaly surface.

Flowers: Separate male and female flowers. Male flowers borne in spikes at ends of branchlets. Female flowers borne in leaf axils.

Fruit: Fruiting body more or less circular in outline, 15 mm in diameter. Almost concealed by hemispherical spongy appendages, which give the plant its common name.

Uses: The seeds are edible. The young leaves are used as a pot-herb, after boiling in two changes of water to remove bitter soluble salts.

Comments: Old man saltbush (*Atriplex nummularia*) could probably be used similarly.

GREY OR WHITE MANGROVE

Scientific Name: *Avicennia marina*

Aboriginal Name: NGURNULU (Broome)

References: 13, 64, 86, 93, 94, 98, 102, 121

Habitat: Sheltered coastal and estuarine areas in southern latitudes.

Size and Form: Shrub or tree, up to 16 m tall. Bark smooth, grey to grey-brown.

Leaves: Broad, tapering towards the tip, 4 – 7 cm long, 2 – 4.4 cm wide. Borne on short stalks on opposite sides of the stem. Upper surface glossy green, underside white or silvery with a covering of tiny dense woolly hairs.

Flowers: Petals yellow or white, flower 1 cm wide. Scented. Borne in small stalked bunches at ends of branches.

Fruit: Capsules oval, pointed at tip, 2 cm long, 1.5 cm wide. Green. When the floating fruit is stranded in a suitable habitat, two valves open to reveal plantlet.

Uses: The fruits are edible after leaching and cooking.

Comments: Many different seeds and nuts are rendered edible by mashing and soaking, before cooking the resulting pulp as a small unleavened loaf in the coals.

CAMPHOR MYRTLE

Scientific Name:	*Baeckea camphorosmae*
Aboriginal Name:	(Not collected)
References:	32, 72, 76
Habitat:	Rocky slopes and range foothills, on patches of heavy soil.
Size and Form:	Sparse spreading shrub, up to 2 m tall. Branching at ground level.
Leaves:	Oblong, 4 – 6 mm long. Flat, bright-green, in four distinct rows, at right angles to stem.
Flowers:	White to pink, five petals, about 15 stamens, 4 – 6 mm wide. Borne on short stalks close to branches on old wood. Usually solitary but can be in clusters of up to four.
Fruit:	Urn-shaped capsules, with three valves, 5 mm wide.
Uses:	Used to treat headaches, by waving the crushed leaves under the nose of the sufferer.
Comments:	Crushed leaves emit a distinctive camphor-like odour, claimed by some to smell like garlic. The leaves were used by settlers to make 'bush tea'.

CANDLESTICK BANKSIA

Scientific Name:	*Banksia attenuata*
Aboriginal Name:	BIARA (south-west Western Australia)
References:	12, 18, 37, 49, 68, 72, 76, 83, 86, 93, 126
Habitat:	Usually heath or woodland. Often in mixed stands with other banksias.
Size and Form:	Small or medium-sized tree, 4 – 14 m tall. Bark rough, dark-grey. Younger branches downy.
Leaves:	Narrow, tapering towards the base or wedge-shaped, 8 – 15 cm long, less than 1 cm wide. Flat, stiff, evenly serrated.
Flowers:	Numerous terminal cylindrical flower spikes, erect and often slender, up to 25 cm long. Intense sulphur-yellow.
Fruit:	Long narrow cylindrical cones, covered with dense grey furry hairs, containing prominent egg-shaped or globular capsules.
Uses:	The branches and bark are used as a 'touchwood' to preserve fire embers. Old flower spikes are used as tinder in fire-making. Nectar is sucked directly from the flower, or flowers are soaked in water to make a sweet drink.
Comments:	Called a 'honeysuckle' by early settlers because of its abundant nectar.

PROPELLER BANKSIA

Scientific Name:	*Banksia candolleana*
Aboriginal Name:	(Not collected)
References:	9, 37, 49, 108
Habitat:	Coastal sandplain and heath communities.
Size and Form:	Multi-stemmed 'mallee' shrub, up to 1 m tall. (If cultivated, can grow to tree 5 m tall.) Branching from a large woody underground structure known as a lignotuber.
Leaves:	Elongated and narrow, divided to midrib with acute triangular lobes, up to 40 cm long, 20 mm wide. Grey-green beneath, bright green above.
Flowers:	Inflorescence golden (rarely light-yellow), cylindrical to globular, to 14 cm and 8 cm wide.
Fruit:	Rarely single, normally fused, usually with 2 – 4 follicles to 6 cm long and 3 cm wide, spread around a central spike in imitation of a marine propeller. Hence common name.
Uses:	The flowers are a rich source of nectar.
Comments:	Between April and early August, nectar-sucking was a great social occasion for Aboriginal groups in the Badgingarra and Eneabba areas. There is anecdotal evidence that they became 'drunk' when the large quantities of nectar fermented in their stomachs.

BULL BANKSIA

Scientific Name: *Banksia grandis*

Aboriginal Name: PULGARLA (south-west Western Australia)

References: 12, 18, 37, 49, 68, 76, 83, 86, 93

Habitat: Widespread and common throughout many habitats in its range. Stunted form on granite rocks and limestone cliffs of south and west coasts.

Size and Form: Tree, up to 8 m tall. Rough grey bark. Twisted or gnarled trunk.

Leaves: Shining, dark-green, 20 – 50 cm long. Regular triangular incisions in to the pale central vein.

Flowers: Cylindrical golden-yellow candles, 25 – 30 cm long, 10 – 12 cm wide. Standing erect above the foliage.

Fruit: Cones large, oblong to cylindrical, with narrow dome at apex where seed capsules have not formed.

Uses: More than a tablespoon of nectar can be obtained from each flower, either by sucking the flower, or by soaking it in water.

ROUND-FRUITED BANKSIA

Scientific Name: *Banksia sphaerocarpa*

Aboriginal Name: NUGOO (south-west Western Australia)

References: 12, 18, 37, 49, 72, 83, 86, 93, 126

Habitat: Sandy heathlands and plains.

Size and Form: Low bushy spreading shrub, 1 – 2 m tall.

Leaves: Needle-like, varying in size and colour in different forms.

Flowers: Globular 'brushes', usually 5 – 8 cm in diameter, on short lateral branches. Yellow or bronze-yellow. Flowering usually spring and summer.

Fruit: Cones, larger than flower spikes and studded with prominent flat-topped capsules with a lip along the join between the two halves.

Uses: Nectar can be sucked from the flowers on wet days, or soaked from the flowers at other times.

Comments: To obtain the nectar, a large number of the flowers were soaked in water held in a skin cloak, which was placed over a hole scooped in the ground.

FRESHWATER MANGROVE

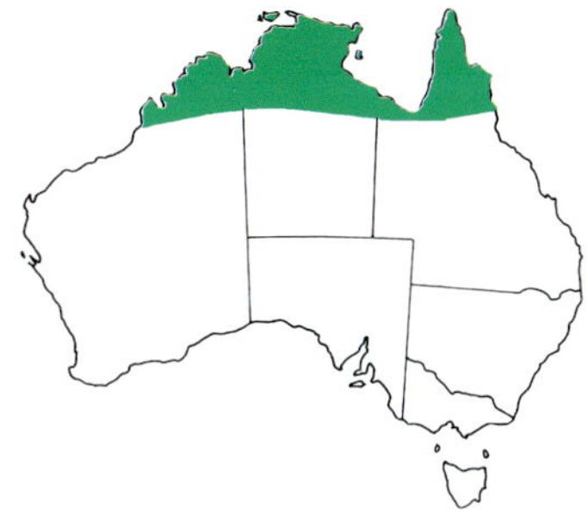

Scientific Name: *Barringtonia acutangula*

Aboriginal name: MADJALA (Fitzroy Crossing)

References: 10, 12, 102, 112

Habitat: Edges of freshwater creeks and lagoons, and of swamps near coastal plains.

Size and Form: Spreading robust tree, up to 10 m tall.

Leaves: Varying in shape, 6 – 14 cm long, 2.5 – 4.5 cm wide, margins usually slightly serrated.

Flowers: Pendulous red flower-heads, 1 cm in diameter, with numerous stamens.

Fruit: Almond-shaped, 3 – 5 cm long, with four prominent 'wings' or lobes.

Uses: The bark is pulped and used to stupefy fish, but there is no evidence of poisoning of livestock. The bark is also boiled in water and the liquid used to treat earache and catfish stings.

WESTERN BLUEBELL

Scientific Name: *Billardiera variifolia*

Aboriginal name: KURUP (south-west Western Australia)

References: 32, 71

Habitat: Sandy soils on heathlands and forest margins.

Size and Form: Shrubby or slender climber, 1 – 3 m tall. Young shoots slender and hairy.

Leaves: Broad, tapering towards tip, 1 – 2 cm long, 0.3 – 0.5 cm wide. Often deeply toothed or lobed. Dark-green, hairy, with crinkled margins.

Flowers: Dark blue, cup-shaped, 1 – 1.2 cm wide, spreading petals. Sepals long, narrow, pointed and hairy. Flowering January – May.

Fruit: Cylindrical berries, up to 2 cm long, with tiny point at end. Green-purple, white and soft when ripe.

Uses: The fruit is edible and is an important seasonal item in the Aboriginal diet. It was used in jam-making by early settlers.

Comments: The unripe fruit is very astringent. Ripeness, for this and other *Billardiera* spp. fruits, is best tested by gently squeezing the fruit. It is ripe when very soft and pulpy.

TARVINE

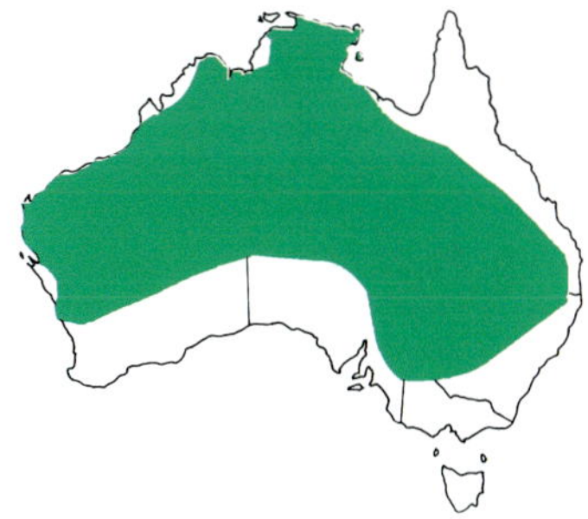

Scientific Name: *Boerhavia coccinea*

Aboriginal Name: URUBA (central Australia)

References: 38, 63

Habitat: Sandplains behind beaches, clearings in sand dune scrub.

Size and Form: Small, sprawling, 1.5 – 2 m in diameter. Stems reddish, slightly hairy.

Leaves: Composed of three rounded leaflets, up to 2 cm wide. Often appears almost leafless.

Flowers: Pink, tiny.

Fruit: The inconspicuous fruit structure is enclosed within the base of the flower parts.

Uses: The main fleshy root can be roasted in the fire 'bush-oven' style, and eaten.

Comments: There are several species of this plant distributed across the continent. They are considered 'weeds'.

NORTHERN KURRAJONG

Scientific Name: *Brachychition diversifolius*

Aboriginal Name: WULAWULANG (Kununurra)

References: 24, 63, 113

Habitat: Mainly open forest.

Size and Form: Tree, seldom taller than 6 m but can reach 25 m in sheltered creek lines. Deciduous; leaves shed late in the dry season.

Leaves: On stems 3 – 10 cm long. Triangular blades, apex ending in long sharp point, 18 cm long, 10 cm wide.

Flowers: Bell-like, pendulous, 5 mm long. Outside green-yellow, hairy. Inside cream with rust-brown spots.

Fruit: Large irregularly shaped pods, 10 cm long, 4.5 cm wide. Opening along one side to reveal 10 – 20 hairy yellow-cream seeds.

Uses: The green sticks with leaves are used as spoons for honey. The seeds are edible and may be roasted to remove hairs before grinding to flour. Inner bark fibres used to make rope. The wood is used for dishes.

Comments: The roots of kurrajongs are an important emergency water source.

DESERT KURRAJONG

Scientific Name:	*Brachychiton gregorii*
Aboriginal Name:	NGALTA (central Australia)
References:	32, 38, 51, 63
Habitat:	Scrubland and sand dune country.
Size and Form:	Single-trunked densely crowned tree, up to 6 m tall. Bark brown, vertically furrowed.
Leaves:	Smooth, shiny, bright-green, 8 – 14 cm long. Deeply divided into three, rarely five, 'fingers' or lobes.
Flowers:	Bell-shaped, 20 mm long, cream with inner reddish markings.
Fruit:	Woody follicles, 4 – 5 cm long. Splitting longitudinally to reveal up to 12 seeds surrounded by shining prickles.
Uses:	The seeds produce a passable coffee substitute when roasted and crushed. Lightly burnt seeds can be ground to make flour and cooked as a damper. Inner bark produces strong fibre for string making. The roots contain water.
Comments:	Seeds should be removed from follicles with a small stick. They are winnowed by pouring between two containers, allowing the wind to carry away the prickly hairs.

LARGE-LEAF KURRAJONG

Scientific Name: *Brachychiton tuberculatus*

Aboriginal Name: (Not collected)

References: 24, 87, 113

Habitat: Mainly in medium-rainfall to high-rainfall areas. Common on margins of clay areas.

Size and Form: Deciduous tree, seldom above 3 m tall, occasionally to 7 m.

Leaves: Almost circular, 20 cm wide, hairy, often with folds. Borne close to the main trunk.

Flowers: Bell-shaped, pink to deep-red, about 3 cm wide.

Fruit: Green woody pods. Borne close to the main trunk and branches. Becoming dark-brown to black when mature, splitting open lengthwise to reveal yellowish seeds. Seeds surrounded with fine hairs, which irritate if touched.

Uses: The gum is used as a sugar substitute. The roots contain water.

Comments: A passable coffee substitute can be made by roasting the seeds and grinding them between two stones. Formerly known as *Sterculia tuberculata.*

RIBBON PEA

Scientific Name: *Brachysema aphyllum*

Aboriginal Name: (Not collected)

References: 32

Habitat: Sandy to loamy soils, especially along watercourses.

Size and Form: Prostrate or dwarf leafless shrub, up to 0.3 m tall, 1 – 2 m wide. Branches up to 1 cm wide, flattened, or winged and notched.

Leaves: None.

Flowers: 'Peaflower'-shaped, bright-red, up to 5 cm long, borne on stalks off branches. Sepals up to 2.5 cm long, green.

Fruit: More or less cylindrical pods, up to 5 cm long, containing many seeds.

Uses: Nectar can be sucked from the flowers.

Comments: Honey-eating birds are attracted to the flowers. It is used as an ornamental plant. The name aphyllum means 'without leaves'.

GREEN PLUM

Scientific Name:	*Buchanania obovata*
Aboriginal Name:	DJAWAD (Derby)
References:	24, 32, 92, 102, 126
Habitat:	Open forest or sandplains behind beaches.
Size and Form:	Small to medium-sized straggly tree, 10 – 15 m tall.
Leaves:	Narrow base, apex broadly rounded, up to 25 cm long, 8 cm wide. Grey-green, with prominent veins.
Flowers:	Cream, 5 mm wide, held in large sprays.
Fruit:	Small, flattened, slightly hairy, yellow-green when ripe, drooping.
Uses:	Toothache is treated by plugging the painful tooth with the shredded wood. A decoction of the inner bark is used to treat sore eyes or as a mouthwash to treat toothache. The fruit is edible, sweet and tasty when soft.
Comments:	*Buchanania obovata* cannot be satisfactorilly separated from a similar species *B. arborescens* until flowering material is obtained.

MILKMAIDS

Scientific Name:	*Burchardia umbellata*
Aboriginal Name:	CARA (south-west Western Australia)
References:	12, 18, 68, 83, 86, 93
Habitat:	Dry woodland, forest and scrub-heath.
Size and Form:	Herbaceous perennial. Arising from 3 – 10 root tubers, each 5 – 10 cm long, about 7 cm below ground.
Leaves:	Two or three shiny grass-like leaves, purple at base, 10 – 30 cm long, 0.5 cm wide.
Flowers:	Flower heads 20 – 30 cm tall, 4 – 10 flowers arising in cluster from top of stalk. White, maturing to pink with darker 'nut-like' centre. Flowering early spring.
Fruit:	Tough three-angled capsule, containing 10 – 20 brown angular seeds.
Uses:	The root tubers are eaten raw or roasted by Aborigines.
Comments:	*Burchardia multiflora*, which is found on granite outcrops, swamps and seasonally wet areas, can also be used in this way.

PALE GRASS LILY

Scientific Name: *Caesia parviflora*

Aboriginal Name: KARR (south-west Western Australia)

References: 72, 83

Habitat: Granite outcrops, swamps woodlands and forests.

Size and Form: Grass-like herbaceous perennial, 15 – 30 cm tall. Numerous erect flowering stems, 20 – 30 cm tall. Roots terminate in swollen oval tubers.

Leaves: Linear, 15 – 30 cm long, erect, margins smooth, suffused with red.

Flowers: Erect compound flower-heads. Flowers composed of six white to mauve 'petals'. Open only during daylight, 4 – 6 mm wide when open. Remaining attached to plant after flowering, but becoming spirally twisted.

Fruit: Capsule two or three-lobed, rarely four-lobed. Splitting to release 2 – 4 spherical black shiny seeds.

Uses: The tubers are eaten raw.

Comments: The Aboriginal name for this plant is tentative, as there is some difficulty in separating references to *Caesia* spp. and *Burchardia* spp.

WHITE CYPRESS PINE

Scientific Name: *Callitris columellaris*

Aboriginal Name: WANGARDI (Yuendumu area)

References: 28, 38, 51, 63, 87

Habitat: Light soils, sometimes near the coast but common throughout arid inland regions.

Size and Form: Medium to large tree, 10 – 30 m tall. Evergreen, resinous, aromatic. Dark-grey rough bark.

Leaves: Small, scale-like, in groups of three or four along the branchlets.

Flowers: Separate male and female flowers. Female structures are small cones with six scales, borne on branchlets. They mature as branches grow and persist until fire or drought. Male structures are tiny deciduous cones.

Fruit: Fruiting cones dark-brown, globular, 2 cm in diameter, with hard woody scales.

Uses: The leaves and twigs are boiled and the liquid dabbed on rashes or used to bathe the chest for treatment of colds. Used in 'smoking' for general illness. The gum is used as a wax or cement after adding ash as a temper.

Comments: 'Smoking' involves building a fire rich in coals, smothering it with numerous branches of the selected plant, and immersing the patient in the ensuing thick white smoke.

NORTHERN CYPRESS PINE

Scientific Name: *Callitris intratropica*

Aboriginal Name: GUDI-GUDI (Broome)

References: 63, 98, 113

Habitat: Gravelly soils and rocky hillsides.

Size and Form: Tree, up to 12 m, rarely 18 m tall. Bark rough, grey, fissured.

Leaves: Leaflets tiny, closely pressed to stems.

Flowers: Separate male and female cones, like other conifers.

Fruit: Cylindrical male cones on branch ends. Conspicuous round female cones on short side-branches.

Uses: The tender pink inner bark is crushed, infused in water, and applied to the stomach to treat diarrhoea. The strong smell is supposed to repel sharks, and the aromatic wood is sometimes shaved onto the water to achieve this.

Comments: Paddles for log rafts are cut from this timber, but the naturally paddle-shaped trunks of *Camptostemon schultzii* were probably used more commonly before the arrival of European paddle models.

TURKEY BUSH

Scientific Name: *Calytrix exstipulata*

Aboriginal Name: MANGADA (Kalumburu)

References: 24, 87

Habitat: Sandy soils that may become swampy in wet season.

Size and Form: Upright straggly shrub, up to 2.5 m tall.

Leaves: Tiny, needle-like, lying close to stem.

Flowers: Numerous narrow pointed mauve-pink petals. About 1 cm in diameter.

Fruit: Seeds long, thin, pointed, with persistent star-shaped calyx.

Uses: The leaves are crushed in the hand to release oils, then applied to the body as a liniment for aches and pains. The wood is used for digging sticks and small batons for stone-flaking. Such batons are used in the 'pressure' technique, where tiny flakes of stone are forced from an artefact by means of pressure, rather than by percussion as in the 'knapping' technique.

Comments: The hard dense wood is good firewood.

KAPOK MANGROVE

Scientific Name: *Camptostemon schultzii*

Aboriginal Name: TJULBUL (Broome area)

References: 32, 98, 113

Habitat: Common in coastal areas, often in saline conditions.

Size and Form: Tall shrub or small tree, 5 – 9 m tall, 3 – 6 m wide. Spreading branches with angular branchlets.

Leaves: Oval, blunt, 5 – 8 cm long, 1.5 – 2 cm wide. Leathery, shiny upper surface, underside covered in dense rounded scales. Stalks 1.5 – 2 cm long.

Flowers: Composed of five lobes, each 6 mm long. Borne in groups of up to three on small branchlets behind the new leaf growth.

Fruit: Capsules containing two woolly seeds.

Uses: The wood can be charred in a fire, softened in sea water and applied to sores once or twice a day. The wood is light, and is used to made double rafts (KALWA, Broome) in the Kimberley region of Western Australia.

MANGO BARK

Scientific Name:	*Canarium australianum*
Aboriginal Name:	WANDA (Kalumburu)
References:	24, 32, 102
Habitat:	Jungle, open forest and on rocky hillsides.
Size and Form:	Tree, up to 12 m tall. In jungle habitat bark pale, almost white, with thin scales. Bark of trunk resinous.
Leaves:	Compound, each composed of 7 – 9 leaflets. Leaflets oval or elliptical, 15 cm long, 5 cm wide, hairy, strongly ribbed. Shed during dry season.
Flowers:	Composed of a three-lobed calyx and a three-petalled corolla. Inconspicuous, in bunches in leaf forks.
Fruit:	Grey-green, hard, succulent, egg-shaped, 2.5 cm long.
Uses:	The fruit is used to stop haemorrhaging after childbirth. An infusion of the resinous bark is drunk for diarrhoea and stomach pains. The seeds are said to be edible if roasted. The gum is used to cement and waterproof bindings on fish hooks. The trunks can be used to make dug-out canoes.
Comments:	The inner bark smells strongly of mango, hence the common name.

NATIVE CURRANT

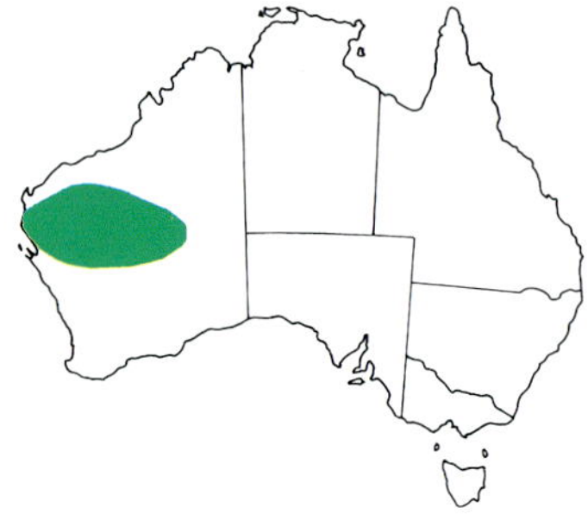

Scientific Name:	*Canthium latifolium*
Aboriginal Name:	NANUNG (Newman)
References:	13, 47, 51, 74, 76, 82, 87, 123 127
Habitat:	Breakaways and mulga-dominated low woodland, often in shade beneath other trees.
Size and Form:	Shrub or small tree, up to 2 m tall, sometimes in dense clumps up to 5 m wide.
Leaves:	Rounded, 2 – 8 cm wide, rigid, leathery, yellow-green. Short stem. Prominent veins, particularly on underside.
Flowers:	Funnel-shaped, 5 – 7 mm long, lobes longer than tube. White, hairy inside, grouped in bunches.
Fruit:	Globular, 4 – 9 mm in diameter. Green, becoming black when ripe, containing one or two black seeds.
Uses:	The fruits are prepared by stewing in water; the skins and seeds are discarded. Eating large amounts of the raw fruit causes minor bleeding of the lips. 'Elbows' cut from the trunk and side-root section are used to make boomerangs.
Comments:	The fruit is abundant only in good seasons. Other *Canthium* spp. exist throughout Australia, and these too have edible fruits.

NIPAN

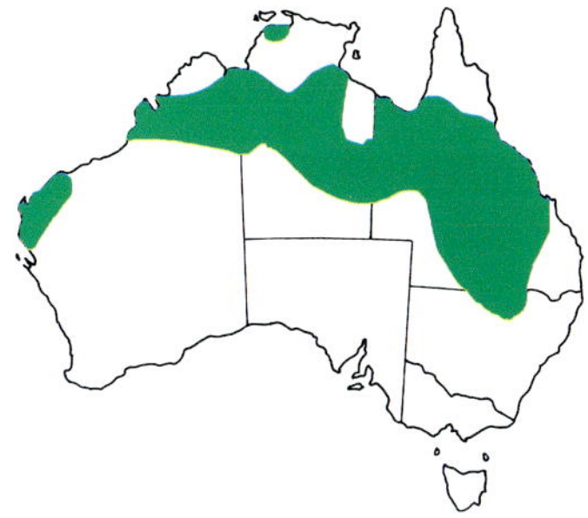

Scientific Name:	*Capparis lasiantha*
Aboriginal Name:	TJIRRWIRLINY (Roebourne area)
References:	51, 60, 63, 102
Habitat:	Throughout northern, tropical, inland Australia on a range of soils. Often partly beneath trees and taller vegetation.
Size and Form:	Sprawling shrub, up to 3 m tall. Whole plant covered with minute rust-coloured hairs, which become greyish on older parts.
Leaves:	Oval or elliptic, 3 – 10 cm long. Thick, corky. Lateral veins prominent on underside of older leaves.
Flowers:	White, petals 1 – 2 cm long. Many stamens. Solitary or in small clusters in leaf axils.
Fruit:	Ovoid. Turning yellow and splitting open when ripe, revealing numerous shiny black seeds.
Uses:	The fruit is edible and succulent. Nectar is used as a cough remedy. An infusion of the whole mashed plant is applied for treatment of swelling and snakebite.
Comments:	The fruits are collected as they begin to turn yellow, and before they split open. They attract ants when ripe.

WILD ORANGE

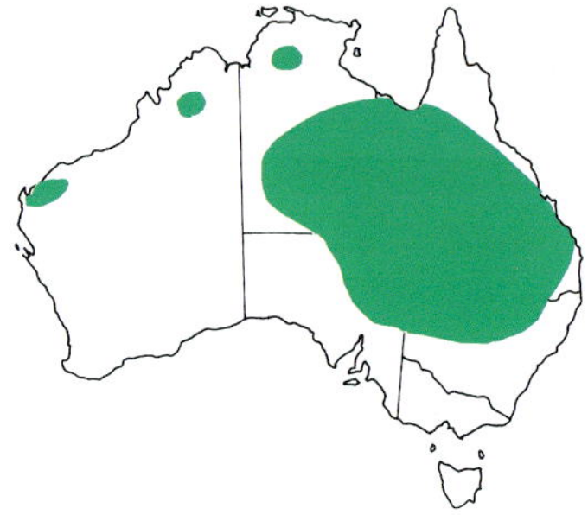

Scientific Name: *Capparis mitchellii*

Aboriginal Name: KATJIWARA (Roebourne)

References: 21, 28, 43, 68

Habitat: Along watercourses and on adjacent plains. Usually on sand or clay-loam.

Size and Form: Bushy shrub to small tree, up to 6 m tall. Dense rounded crown.

Leaves: Long, oval, tapered at base, 2.5 – 3.5 cm long. Young leaves and branches covered with woolly layer of fine hairs.

Flowers: Tightly closed buds on long stalks, opening to reveal four or five white petals and numerous prominent stamens. Flowering in any month.

Fruit: Spherical dull greenish berries, up to 7 cm in diameter when ripe, blackening with age. Containing several seeds, 1 cm long.

Uses: The fruit is somewhat sour, but edible. Unopened buds can be used like 'capers' for flavouring other foods.

Comments: The fine-grained white wood has been used to make tobacco pipes.

FLINDERS ROSE

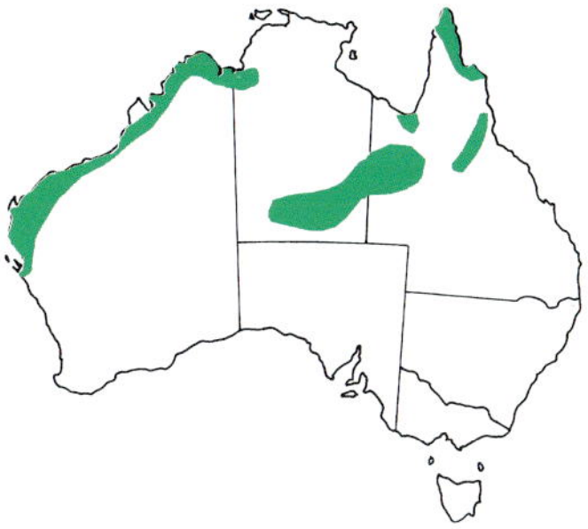

Scientific Name: *Capparis spinosa* var. *nummularia*

Aboriginal Name: PATJULA (Pilbara)

References: 43, 51

Habitat: Coastal plains on sheltered dunes and rocky outcrops, along riverbanks and ephemeral watercourses, around claypans.

Size and Form: Sprawling shrub, 1.5 m tall, up to 4 m wide.

Leaves: Rounded, 2 – 4 cm wide, smooth.

Flowers: Pink-white. Four or five petals almost hidden behind a mass of protruding stamens. Stamens 3 cm long.

Fruit: Ovoid, up to 4 cm long, greenish yellow or red, splitting lengthwise when ripe. Flesh pinkish with many black seeds.

Uses: The flesh is edible and sweet, but the seeds are very hot if cracked, so it is best to leave them intact when sucking the flesh.

Comments: Ants feed on the fruit, and their activity sometimes indicates the location of ripe fruits on the shrub.

BUSH POMEGRANATE

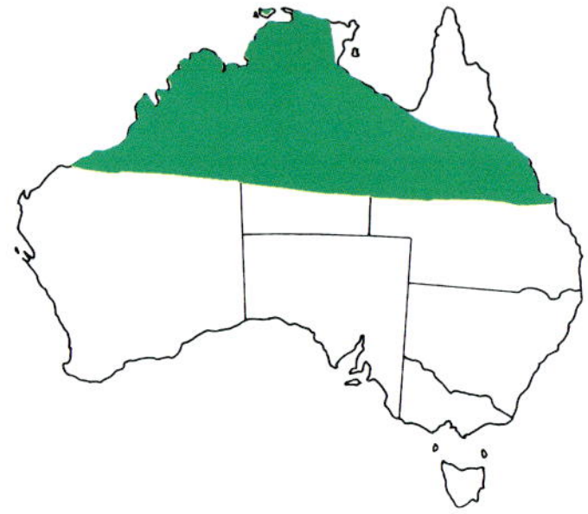

Scientific Name:	*Capparis umbonata*
Aboriginal Name:	DJUGGURUNG (Kununurra)
References:	2, 12, 47, 51, 58, 87, 91, 102
Habitat:	Usually on red sandy soils.
Size and Form:	Small tree, up to 8 m tall, with narrow drooping foliage and rough bark.
Leaves:	Linear with a rounded point, 10 – 25 cm long, 0.7 – 2 cm wide. Narrowing into a stalk 1 – 3 cm long.
Flowers:	Composed of four flat white petals, 2 – 3 cm long, and 50 – 100 long protruding stamens. Outer sepals pointed, 15 mm long. Borne in leaf axils at ends of twigs. Stalk 2 – 5.5 cm long.
Fruit:	Woody, globose, 5 cm in diameter. Yellow or red with pink flesh. Smooth shell that does not open at maturity. Long stalk.
Uses:	The bark is boiled in water, producing a red liquid that is used to bathe aches and pains. The leaves are crushed and steeped in water and the resulting liquid used to bathe eyes.
Comments:	In Timor this species is known as 'rock fruit', because the fruits are so hard to crack open.

BILLABONG TREE

Scientific Name:	*Carallia brachiata*
Aboriginal Name:	(Not collected)
References:	2, 65
Habitat:	In rainforests, near billabongs, along river banks and on the margins of freshwater swamps.
Size and Form:	Tree, up to 15 m tall. Bark cream, thick, fissured and corky.
Leaves:	Oval or broad oval, 14 cm long, 8 cm wide. Papery texture, underside covered with tiny black spots. On stems up to 1.2 cm long.
Flowers:	Small, composed of 5 – 8 greenish cream petals up to 2 mm long. Borne in clusters on short woody stems.
Fruit:	Berries initially green, turning red when ripe in December.
Uses:	The ripe fruit is edible raw.
Comments:	Most plants found in mangrove communities produce fruit that requires some leaching of toxins before being eaten. This species seems to be an exception.

BALLOON VINE

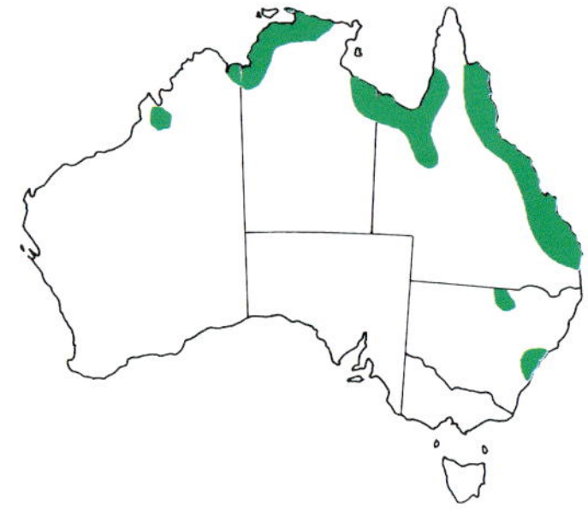

Scientific Name: *Cardiospermum halicacabum*

Aboriginal Name: (Not collected)

References: 25, 32, 46, 54, 96

Habitat: Rocky areas or clay pans.

Size and Form: Straggling vine or climber, 3 – 5 m tall. Stems angular, very hairy.

Leaves: An arrangement of three, three-lobed leaflets. Light-green, thin, hairy. Leaf lobes oval, coarsely toothed, 0.5 – 2 cm wide.

Flowers: Four white petals, corolla 8 mm wide. Borne in groups of three or four.

Fruit: Three-cornered capsules with six prominent ribs, pale green. Becoming papery when ripe. Splitting to release 1 – 3 black seeds, 5 mm in diameter, which are attached to the centre of a papery wing.

Uses: The roots are purported to have medicinal properties in the treatment of haemorrhoids, rheumatism and other nervous disorders. The leaves are eaten and used to treat sore eyes.

Comments: Probably an introduced plant. This plant is used medicinally in India and China, and the dried plant is sold in Chinese herbal stores.

CONKERBERRY

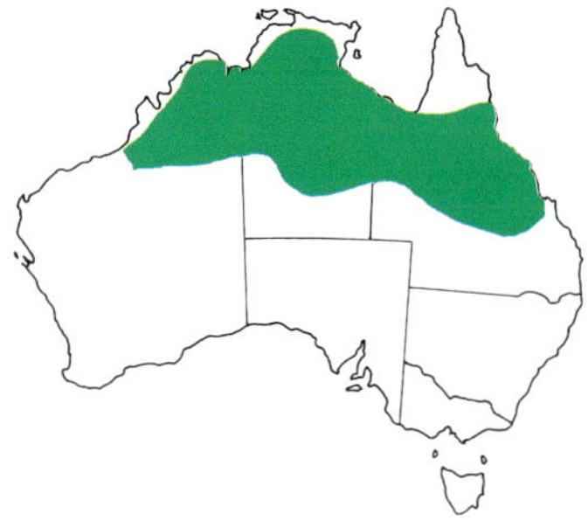

Scientific Name: *Carissa lanceolata*

Aboriginal Name: KUNGGARA (Broome)

References: 25, 32, 72, 113

Habitat: Sand or sandy loam, often forming dense low thickets around the base of isolated trees.

Size and Form: Erect to spreading many-branched low shrub or small tree, up to 3 m tall.

Leaves: Narrow, oval, with rounded tips, 2 – 4 cm long.

Flowers: Small, white, trumpet-shaped, 5 mm long.

Fruit: Oval berries, 5 – 9 mm long. Dull green at first, becoming blackish at maturity.

Uses: The berries are edible when ripe. The dried fruit can be reconstituted by grinding with water. The whole plant can be burnt to repel mosquitoes. Infusions of the roots are used to bathe sores.

Comments: This shrub is a consistent bearer of fruit, even in drought years. It often occurs beneath taller shrubs and trees, where the seeds have been dropped by birds that eat the fruit.

STOUT PIGFACE

Scientific Name:	*Carpobrotus virescens*
Aboriginal Name:	KOLBOGO (south-west Western Australia)
References:	12, 13, 18, 68, 72, 78, 86, 93 126
Habitat:	Coastal sand dunes and arid inland areas.
Size and Form:	Invasive creeping perennial.
Leaves:	Three-sided, up to 8 cm long. Fleshy, rigid, borne on opposite sides of the stem.
Flowers:	Large, showy, daisy-like, many spreading narrow pink petals.
Fruit:	Fleshy, juicy, with many tiny seeds in a thin viscous pulp. Usually red-purple when ripe. Up to 4 cm long, 1.5 cm wide.
Uses:	The fruit is edible raw. The leaves are reputedly eaten steamed or roasted.
Comments:	Fruits gathered from robust plants are less salty then those from plants growing in harsh areas. It was once called the 'hottentot fig', in reference to the small size of the plant and to the fruit, which bears some resemblance to a fig.

A BUSH POTATO

Scientific Name: *Cartonema parviflorum*

Aboriginal Name: WIRIDJUGU (Kalumburu)

References: 2, 24

Habitat: Swamps and damp areas of open forest.

Size and Form: Slender herbaceous perennial, up to 25 cm tall.

Leaves: Long, grass-like, 8 – 15 cm long.

Flowers: Showy yellow three-petalled spikes. Sepals green, hairy, pointed.

Fruit: Small spherical or egg-shaped root tuber, at base of stem, 3 – 5 cm below soil surface.

Uses: The tubers are obtained by pulling up dead plants in the dry season. They are roasted in hot sand and ashes and eaten. When fresh, they have a strong spicy smell.

Comments: The sepals persist after the petals fall, giving the plant a spiky appearance. The tubers of *Cartonema spicatum*, which has shorter flower spikes, can also be eaten after leaching and thorough cooking.

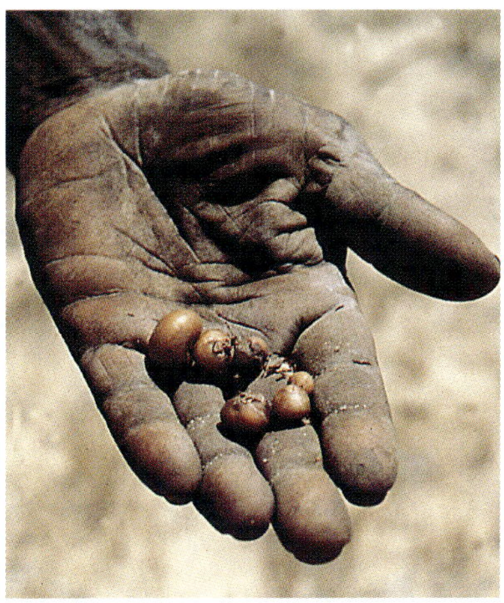

DODDER LAUREL

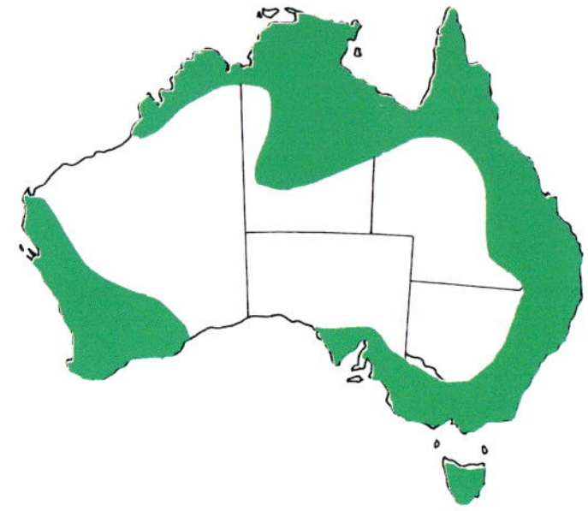

Scientific Name:	*Cassytha filiformis*
Aboriginal Name:	WUNPERRI (Fitzroy Crossing)
References:	2, 12, 13, 51, 65, 102
Habitat:	Sandy and infertile areas, often on heathland and coastal dunes.
Size and Form:	Parasitic scrambling plant, numerous stems, twining over grasses, bushes and trees. Attached to host plant by numerous small oval pad-like suckers.
Leaves:	Indistinct, scale-like, clasping the stem.
Flowers:	Stalkless or stalked spikes. Sepals scale-like, similar to bracts. Petals larger, white, fleshy. Six or nine stamens.
Fruit:	Globular, 8 mm in diameter, containing one seed. Pearl-white when ripe.
Uses:	Edible fruit, which is sweet but insubstantial. Popular with children.
Comments:	Other members of this genus may also produce edible fruit.

A BUSH POTATO

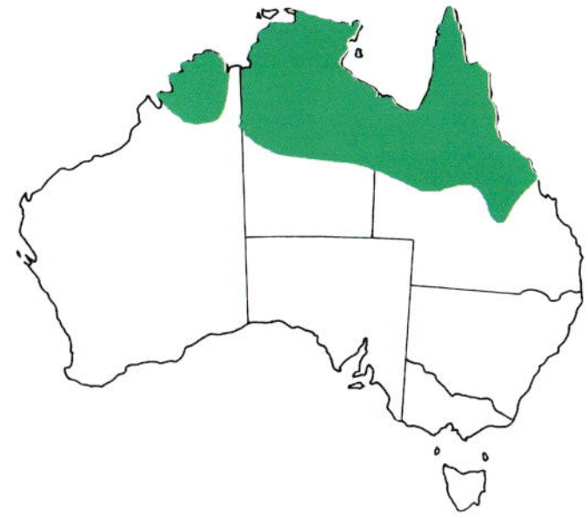

Scientific Name: *Cayratia trifolia*

Aboriginal Name: KARNKAING (Kununurra)

References: 54, 58, 63, 65, 87, 126

Habitat: Widespread and common in lightly forested areas.

Size and Form: Creeper, scrambling through low trees and underbrush.

Leaves: Compound, composed of three leaflets, rounded, serrated, pointed tip.

Flowers: Small and inconspicuous.

Fruit: Inedible black berries in loose grape-like clusters.

Uses: The tuberous root has several globular swellings, 1 – 3 cm wide. From April onwards, these are dug up, cooked and eaten. They are known as 'cheeky' yams, as they cause the mouth and tongue to sting.

Comments: This species is a member of the grape family, but the fruit is not edible.

ENDEAVOUR RIVER VINE

Scientific Name: *Cissus adnata*

Aboriginal Name: (Not collected)

References: 54, 92, 119

Habitat: Along stream banks in open forest.

Size and Form: Vigorous woody scrambling climber, usually shedding all its leaves during dry season.

Leaves: Broad, circular, up to 12 cm long, with shallow-toothed margins.

Flowers: Green or creamy, 2 mm wide, borne in numerous bunches near the ends of branches.

Fruit: Green-yellow ellipsoid berries, 6 mm long. Shiny and black when ripe, resembling grapes.

Uses: Although it is sometimes known as the 'native grape', the fruits of this vine although edible are leathery, and leave an unpleasant taste in the mouth.

PADDY MELON

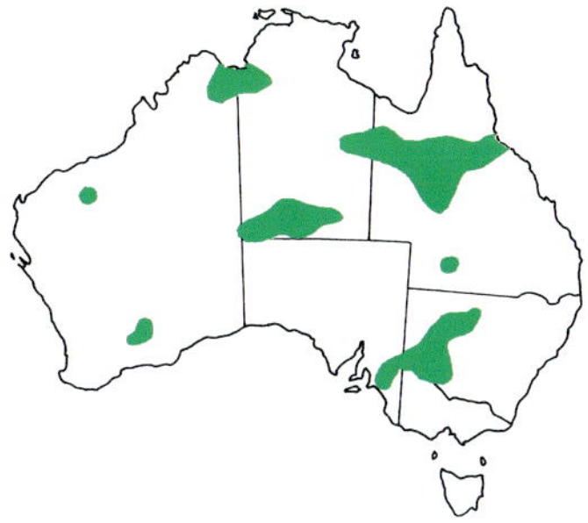

Scientific Name:	*Citrullus colocynthis*
Aboriginal Name:	(Not collected)
References:	28, 51, 63
Habitat:	Widespread and common, particularly in disturbed soil in cultivated areas.
Size and Form:	Trailing herb with perennial rootstock. Spreading horizontally, but growing upwards over other vegetation.
Leaves:	Ovate, with 3 – 5 lobes, up to 10 cm long.
Flowers:	Yellow, 2.5 cm in diameter. Male flowers on stalks arising from leaf axils, female flowers stalkless. Ovary hairy.
Fruit:	Smooth, roughly globular, 5 – 12 cm in diameter. Green mottled with white. Seeds numerous, 5 – 7 mm long, dull yellow-brown.
Uses:	The juice of the fruit, or the boiled fruit is applied to scabies and ringworm.
Comments:	Introduced into Australia. The flesh of the fruit is intensely bitter, even when very ripe. However, South African indigenous peoples have used the fruit as a source of water in times of drought.

SMALL-LEAVED CLEMATIS

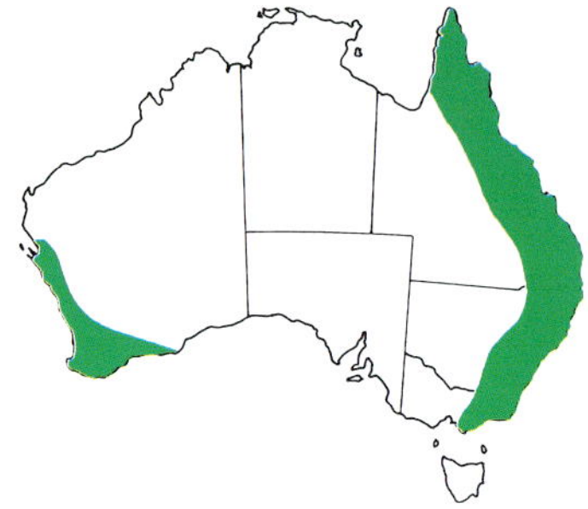

Scientific Name: *Clematis microphylla* var. *microphylla*

Aboriginal Name: (Not collected)

References: 54, 60, 71, 83

Habitat: Widespread and common, particularly in coastal scrub.

Size and Form: Woody climber with wiry stems, up to 5 m long.

Leaves: Dull green, divided into two or three small leaflets with a blunt point.

Flowers: Borne in profusion. Four sepals arranged in a cross shape, up to 3 cm long, tapering towards tip. Petals creamy green, but duller than those of other species.

Fruit: Numerous silky, hairy seeds, each with a feather-like appendage.

Uses: A poultice of the leaves is used to treat itches and rashes, but must be removed after a few minutes to prevent greater irritation.

Comments: A related plant, *Clematis glycinoides,* is called 'headache vine' in Queensland. Inhaling the aroma of the crushed leaves is said to cure headaches.

WESTERN CLEMATIS

Scientific Name:	*Clematis pubescens*
Aboriginal Name:	(Not collected)
References:	128
Habitat:	Confined to karri and jarrah forests.
Size and Form:	Woody perennial climber, up to 3 m tall. Root tubers brown, 10 – 20 cm long.
Leaves:	Composed of three leaflets, 3 – 6 cm long, 1 – 2 cm wide. On long sinuous stalks, which assist in attaching the plant to surrounding vegetation.
Flowers:	White, sometimes tinged pink. Four-petalled, rarely three-petalled. Occurring in bunches, covering the upper part of the plant.
Fruit:	Loose arrangement of single black seeds with long feather-like appendage.
Uses:	The tubers of *Clematis* spp. are said to be edible after roasting and pounding, but they are very hot. The fluff attached to the seeds is useful as a fire-lighting tinder.

TICKWEED

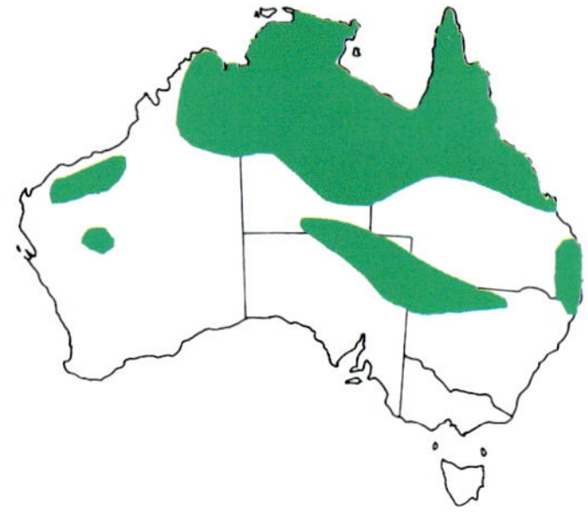

Scientific Name: *Cleome viscosa*

Aboriginal Name: RUJIJIRRUL (Fitzroy Crossing)

References: 12, 28, 60, 63, 65, 97, 102

Habitat: Sandy soils, dry river beds and other deep soils.

Size and Form: Many-stemmed herb, 80 – 90 cm tall, covered with sticky hairs.

Leaves: Composed of 3 – 7 wedge-shaped or spear-shaped leaflets, up to 3 cm long.

Flowers: Yellow, four-petalled, 1 – 2.5 cm in diameter.

Fruit: Long narrow pods, 3 – 10 cm long. Splitting lengthways to reveal numerous small black seeds.

Uses: The leaves are crushed and the vapour inhaled as a decongestant. The leaves are soaked in water, which is used to bathe parts of the body for relief of colds, swelling and rheumatism.

LOLLYBUSH

Scientific Name: *Clerodendrum floribundum*

Aboriginal Name: WITIPI (Balgo)

References: 60, 65, 84, 85, 92, 113

Habitat: Sandy soils in coastal and inland open forests.

Size and Form: Small tree, 3 – 4 m tall. Bark rough, light grey and deeply fissured.

Leaves: Oval, broad at base, tapering at tip, blade 19.5 cm long, 9 cm wide. Distinct network of veins. Borne on stalks 1.5 – 7.5 cm long.

Flowers: White, tubular, long stamens protruding from tube. Star-shaped array of red sepals, up to 12 mm in diameter. Borne in open sprays.

Fruit: Round berries. Initially green, ripening to black. Sepals remain after flower falls, projecting stiffly from base of fruit.

Uses: Eaten during drought; the tissues below the peeled skin of roasted roots being eaten, but not the pith. A decoction of the wood is rubbed on aches and pains. An infusion of inner bark or crushed leaves is drunk as a tonic.

DESERT POPLAR

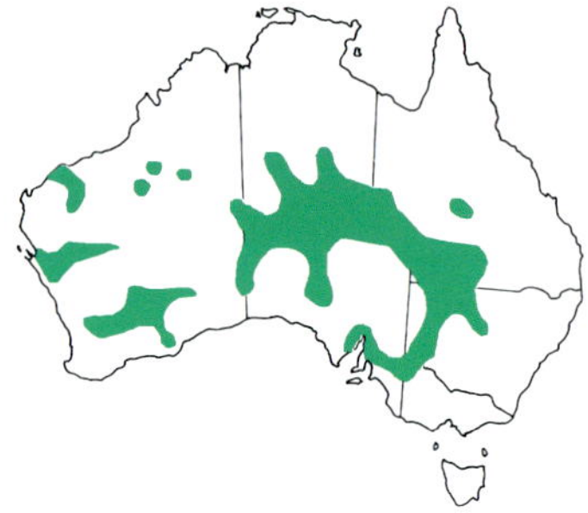

Scientific Name: *Codonocarpus cotinifolius*

Aboriginal Name: KANDURANGU (Meekatharra)

References: 16, 18, 28, 31, 47, 62, 74, 76, 78, 93, 127

Habitat: Red sandy soils in drier inland regions.

Size and Form: Small erect short-lived tree or tall shrub, up to 4 m tall. Smooth pinkish trunk.

Leaves: Broad, oval, tapering to a point, 2 – 5 cm long. Dull, waxy, bright grey-green. Borne on long stalks.

Flowers: Small, insignificant, attached to long stalks near ends of branchlets.

Fruit: Small, bell-shaped, 1 cm long. Comprising 30 – 50 segments, which separate when dry.

Uses: The roots or leaves are chewed as a narcotic, or occasionally made into a 'tea'. Rolled sections of leaf are placed in dental caries to provide relief from toothache. Roots or leaves are powdered with *Acacia cuthbertsonii* and rubbed on aching limbs or teeth.

Comments: Grubs in the roots of this plant must be eviscerated before being eaten, as they are said to contain a 'poison'.

NATIVE TARO

Scientific Name: *Colocasia esculenta*

Aboriginal Name: TJINGU (Derby)

References: 24

Habitat: Widespread and common in seasonally or permanently wet freshwater areas of tropical northern Australia.

Size and Form: Low-growing herbaceous perennial. Stout underground stem, growing at or just below soil level.

Leaves: Ovate, 20 – 50 cm long, 10 – 20 cm wide, held erect on tall stem.

Flowers: Typical of lilies, with the single 'blade' rolled into a cone surrounding the central spike. On stems up to 1.5 m tall.

Fruit: Small fleshy berries, enclosed in the persistent base of the flower parts.

Uses: The tuberous stem is edible. It is cooked overnight in an earth oven, or pounded to a paste, washed and cooked.

Comments: Dies back to underground stem if water levels drop significantly. Improperly prepared, this plant is extremely toxic.

A FLANNEL WEED

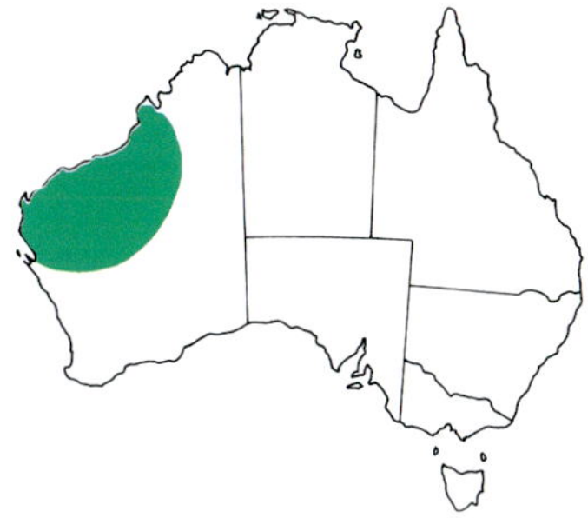

Scientific Name:	*Corchorus walcotti*
Aboriginal Name:	OORUN (Roebourne)
References:	32, 87, 113
Habitat:	Open rocky situations.
Size and Form:	Small spreading shrub, 0.3 – 1.5 m tall. Young shoots densely covered with soft woolly hairs.
Leaves:	Oblong to ovate in outline, 3 – 6 cm long, 1 – 2 cm wide. Thick, grey, covered with dense soft hairs, margins coarsely toothed.
Flowers:	Yellow-orange, 1 – 1.2 cm wide, broad petals, woolly sepals. In axillary clusters of 3 – 6 flowers.
Fruit:	Erect capsules, 1 – 2 cm long. Covered in dense woolly hairs.
Uses:	Bark from the stems is used to make twine.
Comments:	The macerated plant is slimy, and may have medicinal properties. Goannas that have been bitten by snakes are said to bite the leaves of this shrub.

WHITE BUSH APPLE

Scientific Name:	*Cordia subcordata*
Aboriginal Name:	(Not collected)
References:	2
Habitat:	Beaches. Above high tide level in vine thickets.
Size and Form:	Spreading shrub or tree, 2 – 15 m tall.
Leaves:	Broad, ovate, 8 – 20 cm long, 5 – 15 cm wide, rarely two-lobed at base. Fine hairs flanking main veins on lower surface. Long stalks.
Flowers:	Orange. Flower tube up to 1.7 cm long, 5 – 7 broad lobes spreading to 2 cm in diameter. Flower heads composed of 6 – 20 flowers, at ends of branches.
Fruit:	Single-seeded, 2 – 3 cm long, 1.5 – 2.5 cm in diameter, with prominent 'beak' at the tip. Apple-like texture.
Uses:	The fruit is eaten raw.
Comments:	Also known from offshore islands of Western Australia, Malaysia and islands of the west Pacific.

WHITE SAND LILY

Scientific Name: *Crinum angustifolium*

Aboriginal Name: (Not collected)

References: 92, 113

Habitat: Deep sand and sandy loams on river banks, open forest and sandplains behind beaches.

Size and Form: Leaves emerge from ground level and sprawl or stand more or less erect.

Leaves: Broad, strap-like, 20 – 100 cm long. U-shaped in cross-section. Growing from bulbs 8 – 15 cm in diameter, covered in papery remains of old bulb scales.

Flowers: Large, white, trumpet-shaped, stamens purple. Flower heads composed of 3 – 8 flowers.

Fruit: Fleshy globular capsules, 1 – 3 cm in diameter.

Uses: The bulbs are cut into small pieces, soaked and used as an antiseptic for dressing wounds and sores.

Comments: *Crinum* spp. are generally considered to be poisonous, but there are suggestions that this one can be eaten after leaching and cooking.

GREEN BIRD FLOWER

Scientific Name: *Crotalaria cunninghamii*

Aboriginal Name: NGALYIPI (Balgo area)

References: 32, 51, 60

Habitat: Coastal and inland sandy soils, common on sandhill crests in 'spinifex country'.

Size and Form: Erect shrub, up to 2 m tall. Stems pithy, covered with lightly furred yellow-brown bark.

Leaves: Composed of three leaflets. Leaflets ovate, 2.5 – 8 cm long, densely hairy on both sides.

Flowers: Attractive clusters of yellow-green 'peaflowers', 3.5 cm long, darkly veined.

Fruit: Leathery pods, hairy, 3.5 cm long.

Uses: An infusion of the bruised leaves is used to bathe sore eyes and to cure earache. The bark is used to make sandals called BALGA.

Comments: The name of the Perth suburb is not derived from the name of this sandal, but from the local Aboriginal name for the grasstree, *Xanthorrhoea preissii*.

ULCARDO MELON

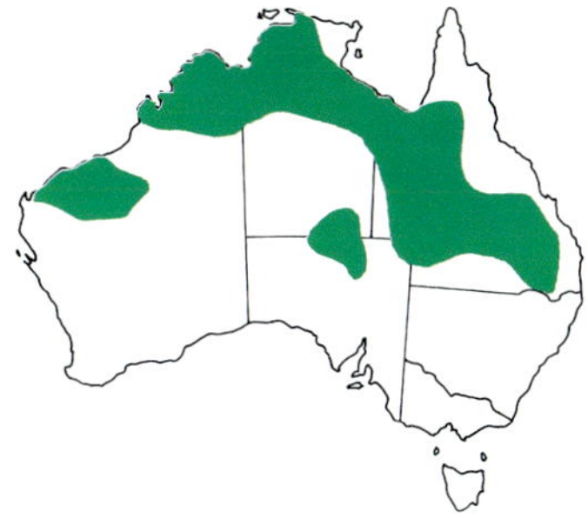

Scientific Name: *Cucumis melo* subsp. *agrestis*

Aboriginal Name: TJIPARTA (Ashburton)

References: 2, 28, 51, 54, 62, 92

Habitat: Usually on grey and heavy clay soils on fringes of watercourses. Grows quickly following rains.

Size and Form: Weak annual vine, climbing with the aid of twining tendrils or trailing across the ground.

Leaves: Roughly oval, with indented base where stalk joins, 2 – 7 cm long. Composed of three or five lobes, middle lobe often dissected into three further lobes. Stalks about as long as leaves.

Flowers: Male flowers 5 – 18 mm long. Petals yellow, 6 – 15 mm long. Tube formed by fusion of sepals, 5 – 6 mm long. Female flowers similar but on shorter stalks.

Fruit: Egg-shaped to oval, 1.5 – 4 cm long, greenish, smooth or hairy. Seeds numerous, pale, 5 – 6 mm long.

Uses: The fruit is edible when fallen from the vine, tasting similar to cucumber.

Comments: An introduced species. The outer skin of the fruit and the vine are very astringent. It is best to squirt pulp into the mouth from the stem end of the fruit without contacting the lips.

SCENT GRASS

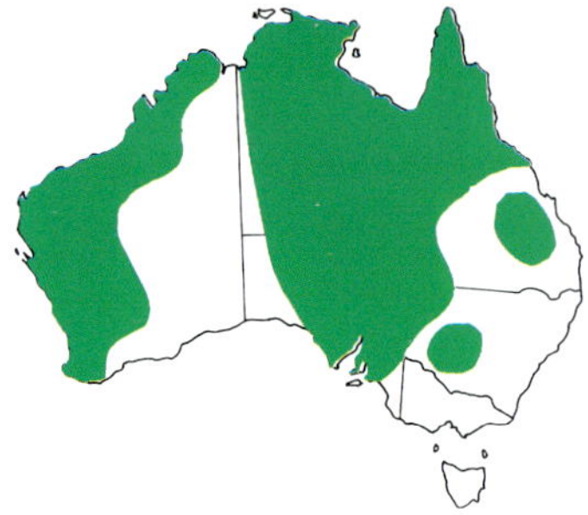

Scientific Name: *Cymbopogon ambiguus*

Aboriginal Name: KALPALPI (Yuendumu)

References: 51, 92, 97, 113

Habitat: Adjacent to rocky outcrops, in crevices along river banks.

Size and Form: Forming tussocks, 30 – 90 cm tall. Citronella-scented when fresh. Flower stems unbranched, smooth.

Leaves: Smooth narrow folded blades, 22 – 45 cm long. Erect. Prominent central vein.

Flowers: Fluffy, typical of grasses. Held in a spike within or above tops of leaves.

Fruit: Seeds with spiky barbed trailing hairs.

Uses: The plant is cut up, boiled in water and the resulting liquid used to bathe the body for general illness. Small quantities can be drunk as a treatment for a sore throat.

Comments: The leaves and bases of the stems are highly aromatic. The strong scent of the crushed leaves and flower spikes give the plant its common name. There are seven *Cymbopogon* species in Australia, all with similar aromatic qualities.

LEMON GRASS

Scientific Name:	*Cymbopogon procerus*
Aboriginal Name:	MARRAKANGU (Halls Creek)
References:	60, 65, 89
Habitat:	Tropical areas. Favours sandy soil in open forest and hilly areas.
Size and Form:	Tall upright grass in isolated tussocks. Thick tough stem, 1 m long.
Leaves:	Broad, rough, wiry, up to 20 cm long, up to 1 cm wide. Often tapering into a sharp point.
Flowers:	Fluffy or silky mass at top of stem. Brown stamens. Fine bristle-like appendages.
Fruit:	Grass-type seed, with fluffy appendages.
Uses:	Used for the treatment of colds. The leaves are chewed or eaten slowly. Sometimes a decoction is drunk, the aromas giving relief to congested lungs and sinuses.
Comments:	The grass is usually lemon or citronella-scented.

NATIVE PEAR

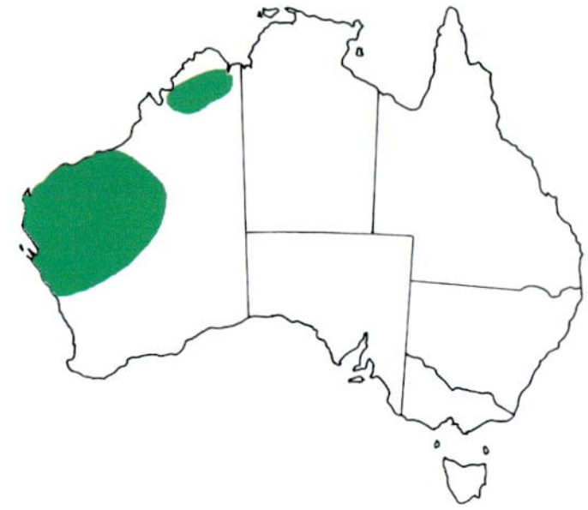

Scientific Name: *Cynanchum floribundum*

Aboriginal Name: DUMARA (Murchison area)

References: 54, 73, 93

Habitat: Tropical regions of northern Australia, generally on sand dunes.

Size and Form: Stout erect herb, up to 1 m tall.

Leaves: Broad, heart-shaped, 2.5 – 5 cm long. Long stalks.

Flowers: Creamy white. Petals in wheel-like arrangement, with five deep lobes. Corona composed of up to 20 lobes.

Fruit: Dry, cigar-shaped or spindle-shaped, winged, 2.5 – 5 cm long. Containing numerous seeds, each with a long silky tuft of hairs.

Uses: The milky sap is rubbed on the body to raise the body temperature during cold weather. Unripe pods are eaten raw or roasted.

Comments: This is one example of a plant containing milky sap that has edible portions. Usually plant exudates of this kind are considered toxic.

NALGOO

Scientific Name: *Cyperus bulbosus*

Aboriginal Name: TJUNTA (Balgo)

References: 13, 47, 73, 74, 113, 127

Habitat: Sandy stream and claypan banks, temporarily wet or inundated areas. Most obvious following good rains.

Size and Form: Herbaceous perennial, up to 40 cm tall. Slender rhizomes.

Leaves: Numerous, elongated, 10 – 25 cm long, 0.3 cm wide.

Flowers: Flower heads composed of many small red-brown florets, 25 mm long, 2.5 mm wide. Arranged in radial clusters at ends of radiating stalks held on a tall stem.

Fruit: Tiny egg-shaped to oval nuts, triangular in cross-section, 1.5 mm long.

Uses: The underground stems grow at a shallow depth and can be dug up by hand. Each terminates in a small tuber resembling a bulb 1 cm wide. These have a sweetish nutty flavour when roasted for a few minutes in hot ashes.

Comments: The tubers of *Cyperus rotundus* are also eaten in this way. This plant is an important food in the desert south of the Fitzroy River. The larger fibrous stems of *Cyperus vaginatus* are used to make string or cord for the manufacture of fishing nets. The stems are flattened with a stone and spun together.

BUTTON GRASS

Scientific Name: *Dactyloctenium radulans*

Aboriginal Name: (Not collected)

References: 32, 51, 87, 93.

Habitat: Sandy soils near rocky hills and along creek beds.

Size and Form: Prostrate annual grass, forming leafy tufts, 15 – 30 cm tall.

Leaves: Flat, tapering towards the apex, 8 cm long. Margins often wrinkled and fringed with long hairs.

Flowers: Flower head composed of 3 – 6 clustered finger-like spikes, 1 – 2 cm long.

Fruit: Small seeds, 1 mm long.

Uses: The seeds are ground to make flour.

Comments: The grass is poisonous to sheep at some times of the year, as it contains hydrocyanic acid.

SPREADING FLAX-LILY

Scientific Name: *Dianella revoluta*

Aboriginal Name: (Not collected)

References: 13, 32, 72, 78

Habitat: Forest and woodland, in the understorey.

Size and Form: Perennial, with underground stems. Forming colonies. Spreading roots fibrous, becoming fleshy and tuberose.

Leaves: Rigid blade with strong ridge in lower half, v-shaped in cross-section, up to 10 cm long. Margins yellow-green, slightly curved. Clustered at base of plant, or evenly spaced along short stem.

Flowers: Deep purple-blue, star-shaped, 1 cm in diameter. Stem up to 1 m long.

Fruit: Blue-purple, egg-shaped, 8 mm long. Borne at ends of stems.

Uses: The fruit is eaten raw. The roots are pounded, then roasted on hot rocks. The fibre contained in the leaves is strong and has a silky texture.

Comments: There are a number of species and subspecies of *Dianella* throughout Australia, which may be used similarly.

SAND SAGE

Scientific Name: *Dicrastylis exsuccosa* (variety unknown)

Aboriginal Name: TILPA (Balgo)

References: 12, 16, 32, 47, 62, 74, 76, 85, 91, 127

Habitat: Deep red sandy soils.

Size and Form: Dwarf or small shrub, 0.3 – 1.5 m tall, 0.3 – 1 m wide. Branches densely covered with golden-yellow hairs.

Leaves: Broad, tapering towards a blunt tip, 3 – 10 cm long, 0.7 – 2.5 cm wide. Thick, densely hairy when young, becoming rough. In pairs, with successive pairs borne at right angles to each other. Aromatic stems and foliage.

Flowers: Tubular, 0.5 cm long, golden-yellow. Clustered in terminal pyramidal flower-heads, 7 cm long. Flower heads borne on branches in the leaf axils. Sepals covered in golden hairs.

Fruit: Oval, 3 mm in diameter, covered with short hairs.

Uses: Plants are beaten in an earth hollow to obtain the fluffy covering of the stems. The fluff is used in making fire, or in body decorations.

Comments: Several other *Dicrastylis* spp. exist, and may be used similarly.

NATIVE YAM

Scientific Name: *Dioscorea hastifolia*

Aboriginal Name: WARRAIN (south-west Western Australia)

References: 10, 12, 13, 47, 72, 76, 77, 83, 93, 124

Habitat: Light fertile soil on lowlands, and amongst basaltic and granite rocks.

Size and Form: Herbaceous perennial climber, up to 2 m tall. Elongated white tuber, 10 – 20 cm long, 1 cm wide, up to 1 m below the surface. Tuber swollen at distal end, covered in numerous small roots.

Leaves: Oblong to oval. In young plants prior to climbing, 1.5 cm long, 1 cm wide. Becoming elongated to 4 cm long, 0.5 cm wide. Borne singly, at different points on the stem.

Flowers: Axillary yellow-cream catkin-like spikes, 3 – 5 cm long. Male plants 18 – 25 flowers per spike; female plants fewer flowers per spike.

Fruit: Seeds circular, suspended between three prominent wings, 2 cm long and 2 cm wide overall.

Uses: The tubers are roasted, pounded and eaten.

Comments: WARRAIN diggings, indicated by hollows, are common along the upper reaches of the Swan River, in the Darling Range of Western Australia.

LONG YAM

Scientific Name: *Dioscorea transversa*

Aboriginal Name: KULLAKA (Arnhem Land)

References: 2, 32, 103, 106, 113

Habitat: Thickets and rainforest. Found in association with *Dioscorea bulbifera*, the round yam.

Size and Form: Twining creeper scrambling through undergrowth. A typical yam, with underground tubers, usually 45 cm long, 6 cm wide.

Leaves: Heart-shaped, with lobes well apart, 10 cm long, 6 cm wide. Bright, shiny, with 5 – 7 veins running the length of the leaf.

Flowers: Female spikes 10 cm long, with numerous small flowers. Flowering male stems commonly have two spikes in each leaf axil.

Fruit: Winged capsules. Width greater than length.

Uses: The tubers, which are cooked in coals and eaten, taste like potato.

Comments: Many of the so-called yams are 'cheeky', meaning hot, and must be sliced and soaked in water before cooking. This plant is reputedly the source of an Aboriginal cure for skin cancer (ref. 32).

YELLOW HOP-BUSH

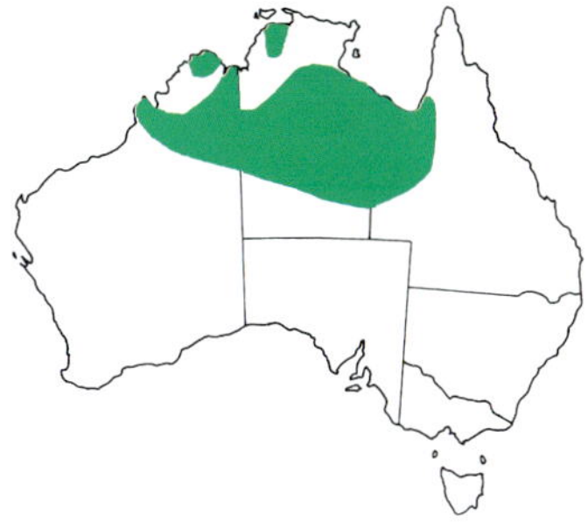

Scientific Name:	*Dodonaea lanceolata* var. *lanceolata*
Aboriginal Name:	PIRRUNGU (Pilbara)
References:	12, 60, 63, 65, 91, 102, 113
Habitat:	Sandy soils on creek banks, in open forest or on well-drained rocky hillsides.
Size and Form:	Spindly shrub, up to 3 m tall.
Leaves:	Narrow to oval, 4 – 12 cm long. Almost veinless, margins straight. On stalks.
Flowers:	Insignificant, green-yellow, without petals.
Fruit:	Distinctive three-winged capsules, 20 mm long and 16 mm wide overall. Papery wings up to 5 mm wide.
Uses:	The leaves are macerated and boiled, and the infusion rubbed onto painful areas. A weak solution can be drunk to relieve pain and snakebite.

STICKY HOPBUSH

Scientific Name:	*Dodonaea viscosa* var. *attenuata*
Aboriginal Name:	TJININPA (Wiluna)
References:	13, 18, 28, 60, 74, 76, 78, 82, 93, 126, 127
Habitat:	Often in gullies and drainage lines on deeper soils.
Size and Form:	Shrub, 3 m or more tall. Thin red-brown bark.
Leaves:	Narrow, oval or oblong, 4 – 10 cm long, 1 cm wide. Toothed margins, young leaves sticky. Borne singly at different heights on the stem.
Flowers:	Inconspicuous, petal-less, greenish. Flowers formed at the tip of a growing axis and further flowers formed on branches below. The colourful fruits are often mistaken for flowers.
Fruit:	Purplish three-winged capsules, 1 cm long, 1.4 cm wide. Dried capsules are bitter and resemble hops.
Uses:	The leaves are chewed, without swallowing the juice, to relieve toothache. Also used in the treatment of stonefish and stingray wounds, although this is not a recommended practice.
Comments:	Leafy branches produce 'clean' smoke, which is used to 'smoke' newborn infants.

LEMON WOOD

Scientific Name: *Dolichandrone heterophylla*

Aboriginal Name: DOHWA (Pilbara)

References: 12, 47, 51, 63, 87, 102

Habitat: Sandy soils and dunes.

Size and Form: Extremely variable. Tree, or rarely shrub, 9 m tall. Rough grey bark, pendulous branches.

Leaves: Borne at the same level but on opposite sides of the stem. Composed of 3 – 13 leaflets, 5 – 10 cm long, with prominent veins.

Flowers: White, trumpet-shaped, 3 – 4 cm long, 3 – 4 cm wide. Closing at night, to re-open the next day.

Fruit: Pods, 10 – 30 cm long, constricted around seeds. Containing winged oblong seeds, 8 mm long, 11 mm wide.

Uses: The green leaves are boiled in water, and the liquid used to bathe the patient for general illness. Straight sticks used to make fire by friction.

ORANGE HONEYPOT

Scientific Name:	*Dryandra fraseri*
Aboriginal Name:	BUTYAK (south-west Western Australia)
References:	12, 18, 32, 72, 76, 83, 93, 126
Habitat:	Gravelly soils.
Size and Form:	Spreading dwarf shrub, up to 1.3 m tall. Branchlets hairy when young, erect or spreading.
Leaves:	Divided to midrib, narrow lobes, 5 – 10 cm long, 1 cm wide. Prickly, glaucous. Curved whitish margins.
Flowers:	Flower heads pinkish orange in bud, opening to bright-yellow, 4.5 – 6.5 cm wide. Borne at ends of branches or on short axillary branches.
Fruit:	Two-winged follicles, held in branch ends, 1 cm long.
Uses:	The flowers are a rich source of nectar, which can be sucked directly from them, or dissolved in water.

COUCH HONEYPOT

Scientific Name: *Dryandra nivea*

Aboriginal Name: BUDJARN (south-west Western Australia)

References: 12, 18, 32, 68, 83, 86, 93, 102, 126

Habitat: Sandy or gravelly soils.

Size and Form: Clumping or spreading dwarf shrub, 0.1 – 1 m tall, 0.6 – 3 m wide. Branches erect or spreading.

Leaves: Fern-like, 15 – 40 cm long, 0.5 – 0.8 cm wide. Can be divided to midrib. Pointed lobes. Smooth dark-green upper surface, whitish underside.

Flowers: Flower heads brown-yellow with pink perianth and green style, 2.5 cm long, 4 cm wide. Borne at the ends of short branchlets. Prominent brown hairy bracts.

Fruit: Flat seed held in centre of papery disc, enclosed between two roughly disc-like wings.

Uses: Nectar is sucked from the flowers as they open.

Comments: There may be several forms of this plant. Another species, *Dryandra quercifolia* (PINGURL, south-west Western Australia) is probably also a source of nectar.

PITURI

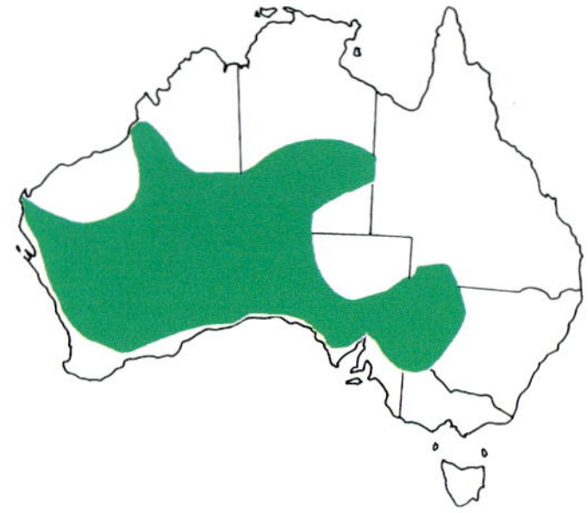

Scientific Name: *Duboisia hopwoodii*

Aboriginal Name: PITURI (central Australia)

References: 13, 16, 47, 72, 74, 76, 78, 83, 93, 126, 127

Habitat: Sandy or gravelly soils.

Size and Form: Usually a shrub, up to 3 m tall. Sometimes a small tree. Bark dark-brown or purplish.

Leaves: Variable in size and shape. Usually tapered at both ends, with straight margins. Borne on short stalks, singly, at different heights on stems.

Flowers: White, five-lobed, bell-shaped, 1 cm long, with purple tube and small green sepals. Borne in open groups at ends of branchlets.

Fruit: Black berries, up to 6 mm long, containing one or two seeds in a dark pulp.

Uses: Dried and powdered leaves and twigs, often mixed with the ash of certain acacias, were chewed and sometimes smoked as a stimulant and narcotic. Used to poison waterholes to incapacitate animals, rendering them easy prey.

Comments: The Aboriginal name is so well known that is used as the common name. Toxic to stock.

GREEN CRUMBWEED

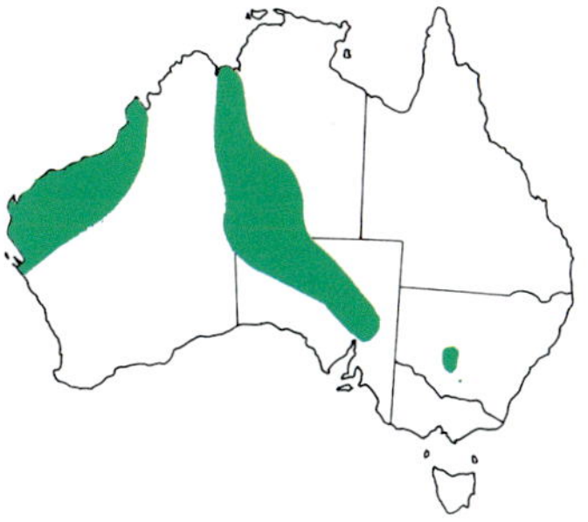

Scientific Name: *Dysphania rhadinostachya*

Aboriginal Name: PILAA (Roebourne)

References: 13, 16, 47, 62, 74, 127.

Habitat: Common in loamy soils on river plains.

Size and Form: Erect herb, up to 30 cm tall, branching from the base.

Leaves: Tiny, 1 – 2 cm long, with undulating margins.

Flowers: Minute. On slender erect green spikes held above foliage.

Fruit: Tiny seeds, 0.7 mm in diameter. Held in old flower parts.

Uses: Whole plants are collected into bundles and scorched with blazing firebrands to open the structures holding the seeds. The plants are then lightly threshed in a large wooden dish, and the residue winnowed from the seeds. The seeds are crushed, mixed with water and cooked in coals. The resulting damper is very sweet.

Comments: *Dysphania plantaginella* (YUMPULA, Pilbara) and *Dysphania kalparri* (KALPARRI, Balgo area) are also used in this way.

WATER CHESTNUT

Scientific Name: *Eleocharis dulcis*

Aboriginal Name: (Not collected)

References: 2, 60, 92, 102

Habitat: Wet places, often near stagnant water, near billabongs and swamps in coastal areas.

Size and Form: Perennial. Rush-like cylindrical tufted stems, up to 1 m tall. Often with roughly ovoid tubers.

Leaves: Leafless.

Flowers: Cylindrical spikelets, 2.5 – 5 cm long, bearing many spirally arranged overlapping scale-like structures. These are 5 – 6.5 mm long, with rounded tips, and conceal the true flower parts.

Fruit: Grey-brown shiny nuts, up to 2 mm long, 2 mm wide.

Uses: A saltwater infusion of the plant is poured into wounds, which are then bound up with the hollow stems of the same plant. Pulp from decaying plants is also bound onto wounds. The tubers are baked and eaten.

Comments: Introduced plant similar to several Australian species. The leaves contain strong alkaloids.

RUBY SALTBUSH

Scientific Name:	*Enchylaena tomentosa* var. *tomentosa*
Aboriginal Name:	(Not collected)
References:	61, 75.
Habitat:	Many soil types, often adjacent to salt pans.
Size and Form:	Straggling shrub, up to 1 m tall.
Leaves:	Slender, more or less cylindrical, 8 – 20 mm long, covered in soft hairs.
Flowers:	Five-lobed perianths with five stamens. Bisexual. Solitary, borne in the angle between leaf and stem.
Fruit:	Berry-like, small, green or bright yellow, turning red when ripe. Often produced in large quantities.
Uses:	The fruit is edible, and can also be used to produce dye.
Comments:	The fruits were traditionally collected by placing a wooden dish beneath the branches and shaking the ripe fruit into it.

NAKED WOOLLYBUTT

Scientific Name:	*Eragrostis eriopoda*
Aboriginal name:	WANGUNU (Balgo area)
References:	13, 16, 42, 62, 72, 74, 76, 91, 123.
Habitat:	Sandy plains and sand dunes.
Size and Form:	Tussock-forming perennial grass, 30 – 50 cm tall, with thickened, woolly base.
Leaves:	Stiff, narrow, with the edges rolled inwards, blade up to 10 cm long. Spreading from stems.
Flowers:	Greenish, with violet tinges. Arranged in open loose panicles, 5 – 15 cm long, up to 8 cm wide. Widely spreading branches holding seed heads.
Fruit:	Oval grains, 0.8 mm long.
Uses:	The seeds are ground into flour, mixed with water and cooked as a damper in the coals.
Comments:	Many of the seeds that are used to make flour are tiny. They are collected by ants, and it is possible to gather them from around ant nests.

NARROW LEAF FUCHSIA

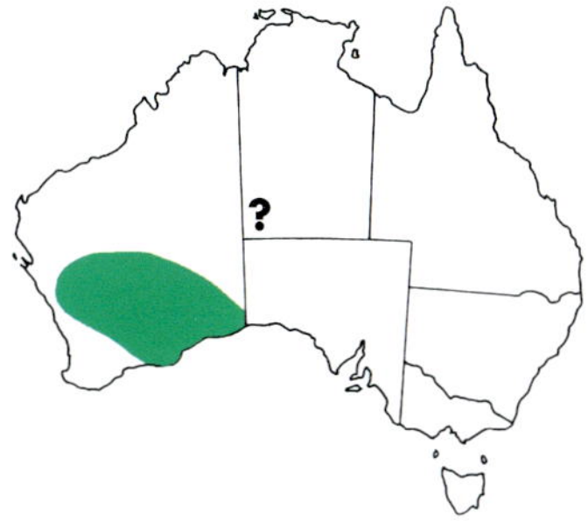

Scientific Name: *Eremophila alternifolia*

Aboriginal Name: YIMI-YIMI (Balgo)

References: 32, 51, 82, 90.

Habitat: Poor and shallow soils on hills and ranges.

Size and Form: Shrub, up to 3 m tall. Young growth sometimes sticky. Branches covered in tiny wart-like growths.

Leaves: Smooth, nearly cylindrical, 10 – 40 mm long, 1 – 1.5 mm wide. Borne singly at different heights on the stems.

Flowers: Tubular, foxglove-like, vivid purple-red, rarely yellow. Sepals purplish, smooth, 6 – 10 mm long. Solitary. Stalks s-shaped, 10 – 30 mm long.

Fruit: Egg-shaped or conical, smooth, 6 mm long.

Uses: Dry leaves, flowers and the bark from young twigs are rubbed into the skin to relieve skin disorders, or used to make an infusion that can be drunk (in small quantities) for colds and sore throats.

Comments: Most *Eremophila* spp. are not grazed by stock. When grazing lands are dominated by them, the country is considered very poor; hence *Eremophila* spp. are commonly called 'poverty bush' in the Murchison/ Ashburton area of Western Australia.

TURPENTINE BUSH

Scientific Name:	*Eremophila fraseri*
Aboriginal Name:	PURRO (Roebourne)
References:	13, 60, 76, 82, 90, 127.
Habitat:	Mulga country.
Size and Form:	Shrub, up to 3 m tall.
Leaves:	Dark-green with shiny or waxy surface, 10 – 40 mm long, 1 – 1.5 mm wide. Borne singly at different heights on the stem.
Flowers:	Tubular or foxglove-like, cream with white spots.
Fruit:	Egg-shaped or conical, smooth, 6 mm long.
Uses:	A decoction of the leaves is drunk to relieve cold symptoms. The plant is also believed to cure toothache and rheumatism. A resinous extract is used as a cement in artefact manufacture.
Comments:	The waxy surface of the leaf (resin) is said to be beaten from the leaves or 'melted out' and formed into flattened paddle shapes on the ends of trimmed twigs for trade and transport.

GEORGINA POISON BUSH

Scientific Name: *Eremophila latrobei*

Aboriginal Name: MARLUPIRNA (Balgo)

References: 13, 16, 47, 74, 76, 78, 90, 91, 97, 127.

Habitat: Sandy clay and stony soils on breakaways and in hummock grasslands.

Size and Form: Greyish densely leafy shrub, 1 – 3 m tall. Branches rough, with swellings from old leaf bases.

Leaves: Very narrow (1 – 5 mm wide) and 15 – 50 mm long. Margins smooth but with the edges rolled inwards, making the leaf somewhat tubular. Borne singly at different heights on the stems.

Flowers: Pale red to purplish, tubular. Solitary. On stalks 1 cm long, smooth or covered in short hairs.

Fruit: Oval, beaked, single-seeded, 5 – 10 mm long.

Uses: The leafy branches are covered with water and boiled for a few minutes. The resulting decoction can be drunk to relieve cold symptoms or used as a topical wash for scabies. Branches are used to 'smoke' babies to relieve bad colds.

Comments: Poisonous to stock.

EMU BUSH

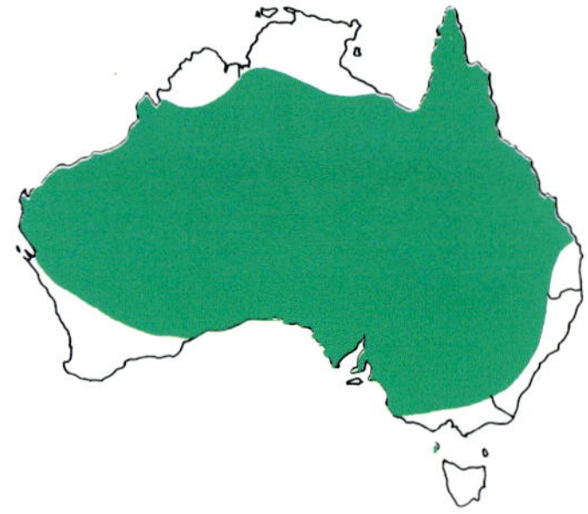

Scientific Name: *Eremophila longifolia*

Aboriginal Name: TULYPURRPA (Balgo)

References: 12, 13, 16, 47, 51, 60, 72, 74, 76, 78, 82, 90, 91, 127.

Habitat: Favours limestone soils. Widespread, but patchy distribution.

Size and Form: Tall erect grey-green shrub, up to 6 m tall, with drooping branches.

Leaves: Narrow and tapered at both ends, 5 – 17 cm long. Borne singly at different heights on the stem.

Flowers: Purplish, 2.5 cm long, hairy on the outside.

Fruit: Egg-shaped or globular. Dark coloured. Fleshy, with a single hard four-celled stone.

Uses: A decoction of the leaves is used to treat sores and drunk to relieve colds. Also used as a 'wash' to bathe the abdomen of women with menstrual pain. The smoke from burning branches is used to cure headaches.

Comments: The seeds of an unidentified *Eremophila* spp. are said to be mashed and mixed with water to make a sweet black drink.

COOKTOWN IRONWOOD

Scientific Name:	*Erythrophleum chlorystachys*
Aboriginal Name:	TJUNTJU (Fitzroy Crossing)
References:	12, 47, 60, 65, 92, 102
Habitat:	Variety of soil types but mainly sandy open forests, often near swamps.
Size and Form:	Tree, up to 30 m tall. Bark dark and fissured on older trees, brown on younger ones.
Leaves:	Compound, composed of 5 – 9 dark-green oval leaflets. Each leaflet 8 cm long, 5 cm wide.
Flowers:	Cream-yellow spikes, 10 cm long. Numerous stamens give spikes a 'bottlebrush' appearance. Individual flowers 1.5 mm long.
Fruit:	Wide flat greenish pods, 20 cm long, 4 cm wide. Ripening to brown.
Uses:	An infusion of the bark is used to treat spear wounds and to rub all over for general malaise. A root infusion is used for cuts and open sores. The plant is sometimes used to poison an enemy. The wood is used to make fighting-sticks or for points and foreshafts of reed or bamboo spears.
Comments:	The leafy branches of this tree are used to make thick white smoke, which is used by Aborigines in ritual purification ceremonies.

MARRI

Scientific Name:	*Eucalyptus calophylla*
Aboriginal Name:	MARRI (Pinjarra)
References:	35, 47, 72, 76, 102
Habitat:	Widespread and common on various lateritic soils.
Size and Form:	Medium to large tree, up to 20 m tall. Bark red-brown, smooth to rough, often cracked into small tile-like sections.
Leaves:	Broad, tapering towards tip, 5 cm wide, 15 cm long. Leathery, with distinct oil glands.
Flowers:	White or pale-pink, up to 5 cm wide, in open clusters.
Fruit:	Woody. Globular to urn-shaped. With internal divisions containing several seeds up to 1 cm long and covered with black skin.
Uses:	Some trunk exudates are considered sweet and palatable. The dried 'blood-like' sap is used as an antiseptic. The seeds are reputedly eaten to cure diarrhoea. The nectar is used to sweeten water, which is much relished.
Comments:	Marri belongs to the group of eucalypts that are called bloodwoods.

RIVER RED GUM

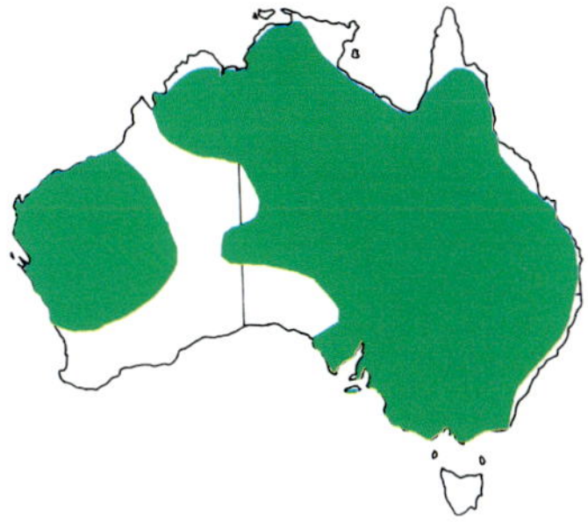

Scientific Name:	*Eucalyptus camaldulensis*
Aboriginal Name:	YALPILINY (Balgo)
References:	10, 12, 13, 35, 47, 49, 58, 60, 74, 76, 78, 82, 83, 87, 91, 93, 102, 127
Habitat:	River banks, alluvial flats and damp low-lying areas subject to flooding.
Size and Form:	Medium to tall tree, up to 20 m tall. Bark deciduous, smooth, grey-white.
Leaves:	Immature leaves slightly bluish, tapered at both ends, 6 – 9 cm long, 2.5 – 4 cm wide, borne at the same level on the axis. Mature leaves borne at different levels on the axis.
Flowers:	Clustered, yellow-cream, up to 2 cm wide. Massed radiating stamens.
Fruit:	Hemispherical, almost globular, 6 – 7 mm in diameter. Broad rim and smooth protruding triangular covers to the interior segments.
Uses:	Fresh young red leaf 'tips' are boiled in water and the strained liquor drunk or used as a 'wash' for colds or coughs. The sap is used as an antiseptic. Leaves are used as a pad or tablecloth on which food is served. Edible grubs are found in certain trees.
Comments:	Flowers nearly all year round. The timber is red coloured.

YATE

Scientific Name: *Eucalyptus cornuta*

Aboriginal Name: YEID (Albany)

References: 18, 19, 35, 47, 68, 86

Habitat: Coastal granites and associated vegetation.

Size and Form: Tree, 15 – 21 m tall. Bark furrowed, dark-grey or black.

Leaves: Mature leaves circular, tapering slightly towards the tip, 8 – 12 cm long, 2.5 cm wide. Conspicuous veins. Borne singly at different heights on the stem.

Flowers: Cream-yellow, 4 cm wide, borne in bunches.

Fruit: Bell-shaped, up to 2.5 cm wide, capped with a long horn-like structure (which encloses the flower parts, but falls as the blossom unfolds).

Uses: The thick purple sap is scraped from stripped sections of bark and mixed with ground acacia seeds to form an edible paste. Bark slabs are used as containers for the mixture.

Comments: The first Western Australian eucalypt to be given a scientific name.

VARIABLE BARKED BLOODWOOD

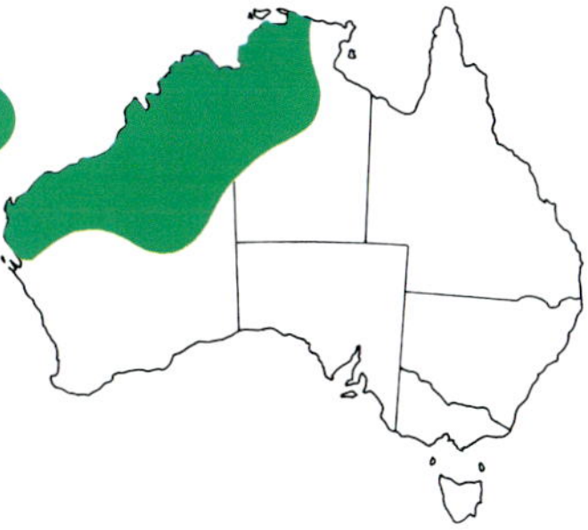

Scientific Name: *Eucalyptus dichromophloia*

Aboriginal Name: PUDANYU (Roebourne)

References: 32, 35, 60, 87, 127

Habitat: Areas with medium to low rainfall, on rocky or shallow sand, often on steep slopes.

Size and Form: Medium-sized tree, up to 12 m tall. Bark smooth, ash-grey, peeling to reveal the inner reddish bark.

Leaves: Narrow, somctimes sickle-shaped, 8 – 20 cm long, 1 – 4 cm wide. Borne singly at different heights on the stem.

Flowers: Flower heads composed of 3 – 6 florets, at ends of branchlets. Stalks 2 – 5 mm long.

Fruit: Globular to slightly urn-shaped, 10 – 13 mm in diameter, thick and firm.

Uses: Red kino is exuded from wounds and natural fissures in the bark and is used as a disinfectant. Kino solution is drunk for colds, bronchial, lung and heart diseases. The nectar is drunk for coughs and colds.

Comments: It is the host tree for an insect that produces galls known as 'bloodwood galls' or 'bush coconuts'. These are the spherical enlarged portions of branchlets, up to 8 cm in diameter. They contain an edible fleshy layer, up to 1 cm thick, with the texture of coconut flesh, and tiny insect larvae, which are also edible.

SPEARWOOD MALLEE

Scientific Name:	*Eucalyptus doratoxylon*
Aboriginal Name:	GEITCHNGUNT (Pinjarra)
References:	18, 35, 68, 70, 76, 86, 126
Habitat:	Rocky places in sheltered heathland, on soils derived from granite, sandstone or quartzite.
Size and Form:	Erect many-stemmed tall shrub or tree, usually 2 – 3 m tall, seldom exceeding 5 m tall. Bark smooth and white.
Leaves:	Narrow, tapering towards the tip, 6 cm long, 1 cm wide. Paler on the under surface. Borne at the same level on opposite sides of the stem.
Flowers:	Small, with slender stalks. The cap covering the bud is sharply beaked and yellow-white, contrasting with the green sepals.
Fruit:	Globular, 8 mm in diameter. Smooth when fresh but appearing wrinkled or striped when dry. Fertile seeds black or grey, sterile seeds red-brown.
Uses:	Aborigines travel long distances to obtain the wood to make spear shafts. The stouter ones are used for digging sticks.
Comments:	Considered a highly decorative species for horticultural uses.

BLUE MALLEE

Scientific Name:	*Eucalyptus gamophylla*
Aboriginal Name:	WARANPA (Balgo area)
References:	47, 62, 74, 76, 84, 91.
Habitat:	Red sand dunes and deep sandplains. Occasionally on rocky hillslopes flanking ranges.
Size and Form:	Usually erect many-stemmed tall shrub, 3 – 5 m tall. Occasionally tree, up to 8 m tall. Bark smooth.
Leaves:	Tapering slightly at tip, 8 cm long, 1.5 cm wide. Young leaves more or less ovate and attractive blue-grey. Borne at same level on opposite sides of stem.
Flowers:	Buds cylindrical to pear-shaped, in leafy clusters. Flowers up to 1.5 cm wide, creamy white. Often in masses.
Fruit:	Pear-shaped, 13 mm long, 7 mm wide. Tapering into very short stalk. Wrinkled when dry. Capsule splitting into short triangular parts when dry.
Uses:	Nectar is sucked directly from the flowers. The seeds are edible, with a nutty taste. Ash from the leaves and bark is mixed with wild tobacco, *Nicotiana* spp., before chewing.

TUART

Scientific Name: *Eucalyptus gomphocephala*

Aboriginal Name: DUOTTA (Cunderdin)

References: 86, 93, 102

Habitat: Restricted to well-drained limestone areas of coastal regions.

Size and Form: Tree, 15 – 35 m tall, with spreading or erect branches. Bark pale, rough.

Leaves: Broadest in the lower half, tapering towards the tip, often sickle-shaped, 10 – 18 cm long, up to 5 cm wide. Pendulous, prominent midrib. Borne singly, at different heights on the axis.

Flowers: Buds capped with rounded cover. Flowers cream, 2 cm wide, occurring in groups of 3 – 7.

Fruit: Bell-shaped, creamy white, smooth or with one faint rib, 2 cm long.

Uses: The bark of the roots is eaten after scorching and pounding. Sometimes this is mixed with gum from *Acacia* spp. Edible grubs can be found in the roots. Hollows between the main trunk and the branches contain water.

Comments: The flowers yield nectar in profusion, which was dissolved in water held in bark dishes.

SNAPPY GUM

Scientific Name:	*Eucalyptus leucophloia*
Aboriginal Name:	MIGUM (Newman)
References:	47, 51
Habitat:	Low stony hills, plateaus and slopes. Sometimes on stony sandplains.
Size and Form:	Tree, 2.5 – 6 m tall. Bark smooth, white, powdery.
Leaves:	Broadest in lower half, tapering towards the tip, 5 – 10 cm long, 1 – 2 cm wide. Yellow-green with faint veins. Stalks 0.8 – 1.5 cm long. Borne singly, at different heights on axis, occasionally at same level.
Flowers:	Composed of 7 – 11 oval or spindle-shaped buds, 5 – 9 mm long, 3 – 5 mm wide. Conical cap covering each bud. Flowers cream.
Fruit:	Hemispherical, woody, 5 – 8 mm long, 4 – 6 mm wide. Seed case splitting into three triangular protruding portions. Seeds oval, yellow.
Uses:	The wood is slow-burning; good for camp fires. Crushed leaves are sniffed to help clear head and lungs. A small parasitic insect builds cases on the leaves in August and September, which are edible and sweet-tasting.

YORK GUM

Scientific Name:	*Eucalyptus loxophleba*
Aboriginal Name:	DOATTA (south-west Western Australia)
References:	12, 13, 18, 55, 68, 72, 76, 78, 83, 86, 93, 126, 127
Habitat:	Sand, sandy loam or loamy soil, in open woodland, on plains and low hills.
Size and Form:	Tree, usually 6 – 12 m or taller. Rough grey-brown bark on main trunk.
Leaves:	Narrow, tapering towards tip, slightly curved with thin point, 7 – 13 cm long. Pale green. Borne singly, at different heights on axis, on stalks 1.5 cm long.
Flowers:	Flower heads composed of 11 conical buds, 2 mm long, 4 mm wide, each covered with a conical cap 5 mm long. Flowers cream and sweetly scented.
Fruit:	Pear-shaped or bell-shaped, 5 – 6 mm long, 4 – 5 mm wide. Opening when ripe to reveal numerous fine brown seeds.
Uses:	The wood is used in the wheelwright trade and in spear-making. Bark of the roots is eaten in the dry season. The stripped bark is wrapped around gum from the stems of *Acacia* spp. and pounded before being eaten.
Comments:	The wood is pale brown, dense and hard.

COOLIBAH

Scientific Name: *Eucalyptus microtheca*

Aboriginal Name: MALANPA (Balgo)

References: 12, 13, 16, 32, 62, 74, 76, 91, 93, 102, 127.

Habitat: Along waterways and seasonally flooded areas.

Size and Form: Tree, up to 12 m tall. Bark varies in different regions, from smooth and white to brownish and rough.

Leaves: Narrow, tapered at both ends, sometimes quite pointed, 6 – 20 cm long. Dull green to slightly bluish. Adult leaves borne singly, at different heights on axis.

Flowers: White, 1 cm wide. Borne in small bunches at the angle between leaf and stem, or at the ends of branchlets.

Fruit: Egg-shaped, 3 – 4 mm in diameter, opening when ripe.

Uses: Saplings are used for spear shafts. Edible grubs are found in the trunk. The inner bark is pulverised and used as a poultice for treatment of snakebites. Lerp scale, a sugary substance exuded by an insect parasitic on the leaves, is gathered from the ground or from around ant nests. This is a sweet and highly favoured food. Water is obtained from root sections.

Comments: Winter flowering. An extremely hard timber.

NORTHERN WOOLLYBUTT

Scientific Name: *Eucalyptus miniata*

Aboriginal Name: MANAWAN (Broome)

References: 92, 98, 102

Habitat: Open forest, on sandstone or quartzite.

Size and Form: Large tree, 7 – 15 m tall. Bark of upper trunk and branches smooth and white, lower bark rough brown and stringy.

Leaves: Narrow, ovate. Mature leaves up to 20 cm long. Dull-green to glaucous, one side paler than other.

Flowers: Bright orange, 3 cm wide. Borne in clusters on short thick stems.

Fruit: Large ribbed gum nuts, up to 5 cm long.

Uses: An infusion made from the bark or leaves is used as a lotion for treating swellings and is taken in sips to cure diarrhoea.

Comments: The blossoms provide native bees with an abundant nectar source from which they produce wild honey. The flowering tree attracts lorikeets, and other nectar-eating birds.

GHOST GUM

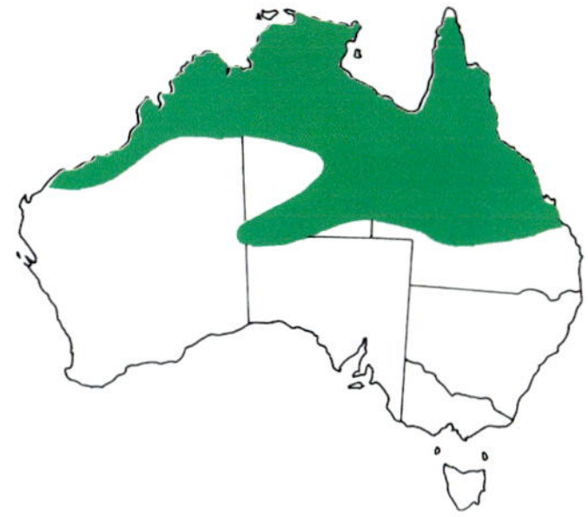

Scientific Name: *Eucalyptus papuana*

Aboriginal Name: (Not collected)

References: 12, 69, 74, 102

Habitat: Usually in sandy soil on rocky hillsides, but also on river flats.

Size and Form: Tree, 15 – 20 m tall. Trunk up to 6 m long and 60 cm diameter. White, satin-like bark.

Leaves: Narrow to broad, tapering towards tip, 13 cm long, 3.5 cm wide. Dull yellow-green to deep shiny green on both surfaces.

Flowers: Creamy, in clusters on short thick stems. Similar to *E. miniata*, but smaller and on shorter stalks.

Fruit: Urn-shaped capsules, 1 cm long, valves inserted deep inside.

Uses: The crystalline or sticky resin is collected, boiled in water until dissolved, concentrated by further boiling and then used to wash sores and cuts.

LONG-FRUITED BLOODWOOD

Scientific Name:	*Eucalyptus polycarpa*
Aboriginal Name:	DJUY (islands off Arnhem Land)
References:	13, 32, 76, 92, 102, 127
Habitat:	Monsoon rainforest, open forest, rocky hillsides and swamps in a variety of soil types.
Size and Form:	Tree, 10 – 12 m tall. Bark grey-brown, thick and scaly.
Leaves:	Broadest in lower half, tapering towards tip, up to 20 cm long, 3 cm wide. Bright green.
Flowers:	White, 2 cm wide, in terminal sprays (corymbs), scented. Buds pale purple and pear-shaped to ovoid.
Fruit:	Ovoid to urn-shaped, truncate, woody. Valves deeply sunken in the capsule and reduced to small teeth. Seeds wing-like.
Uses:	The wood is used for making roasting sticks, digging sticks and other objects. The galls are hammered open and the green grubs inside eaten.

BLACK MARLOCK

Scientific Name:	*Eucalyptus redunca*
Aboriginal Name:	(Not collected)
References:	13, 18, 78, 83, 86, 93, 126.
Habitat:	Savanna woodland on lateritic soils. Only rarely occurring in forest.
Size and Form:	Spreading low mallee shrub or tree with large crown. Up to 6 m tall, 4 m wide.
Leaves:	Young leaves almost triangular, broadest in lower half, tapering towards tip, grey-green. Adult leaves narrower, 8 cm long, 2.5 cm wide, smooth, dull green. Short stalks. Borne singly at different heights on the axis.
Flowers:	In clusters of 4 – 10 buds, short stalks. Sepals forming a cylindrical to bell-shaped tube. Horn-shaped cap covering each bud.
Fruit:	Cup-shaped or cylindrical. Valves composed of sections that come to a fine point and are within or roughly at the rim of the 'cup'.
Uses:	The wood is used for making spears, digging-sticks and clubs.
Comments:	This species is known to the Aborigines as a 'spearwood'.

INLAND BLOODWOOD

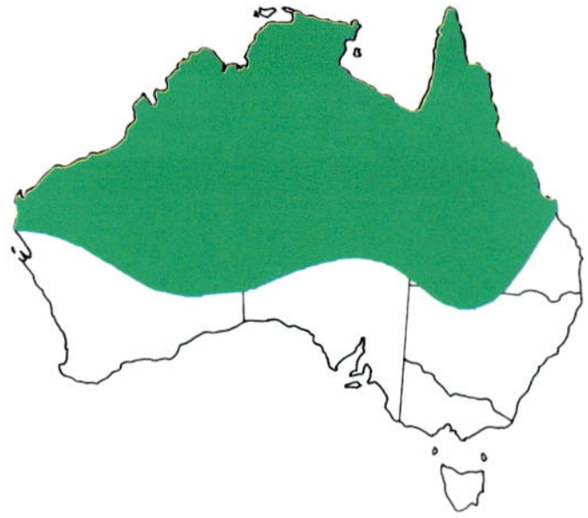

Scientific Name: *Eucalyptus terminalis*

Aboriginal Name: MIJILYPA (western desert)

References: 6, 62, 69, 87, 102, 127

Habitat: Open woodland, river-flats, sandplains and some scree slopes.

Size and Form: Tree, 6 – 11 m tall, often crooked. Bark scaly and tile-like, being divided into small discrete sections.

Leaves: Mature leaves tapering towards tip, up to 10 cm long, dull green.

Flowers: Composed of 3 – 7 cream buds, 8 – 13 mm long, 6 – 9 mm wide.

Fruit: Woody ovoid to pear-shaped capsules, 1.8 – 4 cm long, 1.5 – 2.5 cm wide. Valves are deeply sunken. Seeds yellow-brown.

Uses: The timber is used to make artefacts for ceremonial use. The kino or sap exuded from the trunk is deep-red and may be used as a type of 'varnish' or wood finish.

Comments: This tree is host to a gall-inducing insect which produces 'bush coconuts' (CORABA, Pilbara) on the ends of branches. The non-woody parts of the galls are edible.

DARWIN STRINGYBARK

Scientific Name: *Eucalyptus tetrodonta*

Aboriginal Name: GARDAYNKA (Milingimbi)

References: 32, 92, 102

Habitat: Open forest on a variety of soils. Often in medium-rainfall and high-rainfall areas.

Size and Form: Tree, up to 25 m tall. Bark rough, thick, stringy.

Leaves: Broad, tapering towards tip, to sickle-shaped, with prominent veins. Up to 20 cm long. Dull green on both surfaces. Usually borne singly at different heights on axis.

Flowers: Flower heads composed of three almost stalkless white flowers about 2 cm across.

Fruit: Bell-shaped capsules on short stalks. Smooth or four-ribbed. With persisting 'teeth' of the flowering stage remaining obvious.

Uses: An infusion of the bark and leaves is drunk for aches, pains and coughs, and an infusion of the inner bark is drunk to stop pain and bleeding, especially at childbirth. Sheets of bark are used to make shelters and canoes.

Comments: The persisting teeth on the capsule give the plant its specific name.

WANDOO

Scientific Name: *Eucalyptus wandoo*

Aboriginal Name: WANDOO (south-west Western Australia)

References: 18, 68, 72, 78, 83, 93, 126

Habitat: Sandy loam or gravelly soil with a clay subsoil. Hilly or slightly undulating country.

Size and Form: Tree, up to 21 m tall, occasionally to 30 m. Bark smooth, mottled white and grey, sometimes turning pale salmon pink in autumn.

Leaves: Narrow, tapering to tip, sometimes sickle-shaped, 8 – 12 cm long. Dull grey-green. Borne singly at different heights on the axis.

Flowers: Composed of up to 15 horn-shaped buds, on stalks 2 – 5 mm long. Conical cap covering each bud. Stamens white or creamy.

Fruit: Cylindrical or pear-shaped, 6 – 10 mm long, 5 – 8 mm wide. Valves sometimes protrude slightly from the 'gum-nut'. Seeds brown, round, smooth.

Uses: The flowers produce abundant nectar, making this eucalypt one of the main sources of nectar. It is collected by soaking the flowers in water.

Comments: This tree is mistakenly said to produce large edible tuberous roots called CONNA (south-west Western Australia), which are sweetish and celery flavoured. In fact, it is the country where the CONNA (*Platysace cirrosa*, see p. 202 and ref. 72) grows that is named wandoo country after this tree.

CAUSTIC WEED

Scientific Name: *Euphorbia australis*

Aboriginal Name: PIWI (Pilbara)

References: 13, 16, 47, 74, 76, 83, 127

Habitat: A variety of habitats, mostly on sandy and poor soils.

Size and Form: Many-branched herb, 30 – 60 cm wide.

Leaves: Ovate or nearly circular, 4 – 8 mm long. Asymmetrical or obliquely rounded at base. Almost stalkless, toothed, hairy.

Flowers: Minute, less than 0.5 mm long. Solitary in upper leaf axils.

Fruit: Hairy capsules. Seeds pink, 15 mm long, four-sided, wrinkled between angles.

Uses: The milky sap from the crushed plant has been used for the treatment of skin sores and skin cancer.

Comments: When bruised the euphorbias exude a milky sap, to which many Aboriginal names including NAMANA (from the western desert, meaning breast or milk) refer.

HAIRY CAUSTIC WEED

Scientific Name: *Euphorbia drummondii*

Aboriginal Name: YIPI-YIPI (Balgo)

References: 12, 13, 16, 72, 74, 76, 78, 91, 93, 102, 127

Habitat: Widespread, often as a weed in disturbed areas.

Size and Form: Many-branched herb, up to 60 cm wide, sometimes with red stems arising from a woody rhizome. Stems hairless.

Leaves: Ovate, tip rounded or notched, margins smooth or minutely serrated, 4 – 8 mm long. Borne at same level but on opposite sides of stem.

Flowers: Flower heads composed of small unisexual flowers growing on short stalks out of upper leaf axils. Petal-like parts pink or white.

Fruit: Smooth capsules, less than 2 mm long. Seeds four-sided, pink or reddish-brown, angular, with smooth or wrinkled faces.

Uses: The herb is boiled, and the liquid patted onto scabies, or vigorously rubbed in for pains in the chest. The sap is an irritant, especially of the eyes, and should not be drunk or applied to soft body tissues.

Comments: The bark of *Euphorbia coghlanii* (NAMANA or PIWIRARTHA, western desert words meaning breast or milk), a similar plant that exudes a milky sap, is peeled off the branches and boiled with a little water. The solution is used to treat sore eyes.

LEAFLESS BALLART

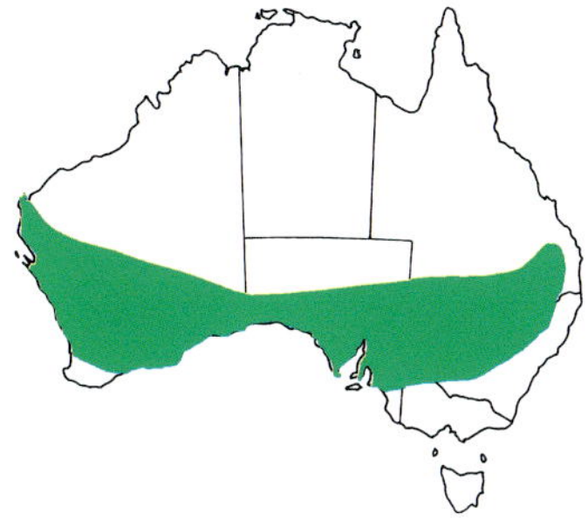

Scientific Name: *Exocarpos aphyllus*

Aboriginal Name: MIRNIKUYAN (Laverton)

References: 16, 18, 72, 76, 78, 82, 83, 93, 126

Habitat: Woodland communities on sandy loams, but also on well-drained clays.

Size and Form: Large untidy shrub or small bushy tree, up to 4 m tall. Often yellow-green in colour.

Leaves: Tiny, scale-like. Soon shed, leaving the rigid branchlets bare.

Flowers: Minute, stalkless, attached to a common stalk, forming a spike.

Fruit: Spherical nuts, 4 – 5 mm in diameter. Attached to a swollen, red, fleshy, reputedly edible berry-like stalk.

Uses: A decoction of mashed stems is taken for colds and used to wash sores. Also used as a poultice on the chest for 'wasting diseases'.

Comments: Parasitic upon the roots of nearby plants.

BROOM BALLART

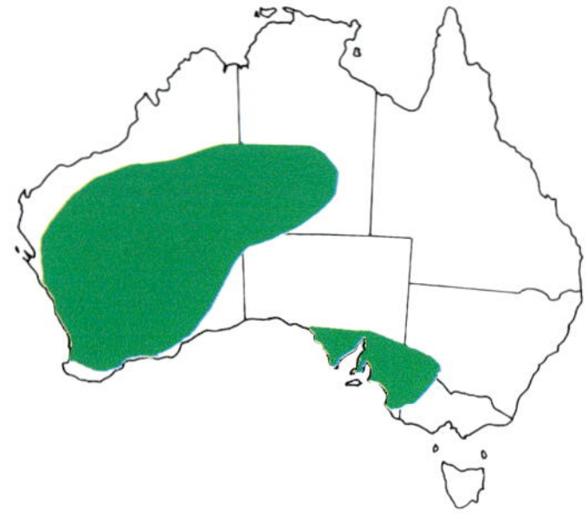

Scientific Name: *Exocarpos sparteus*

Aboriginal Name: CHUCK (south-west Western Australia)

References: 13, 18, 47, 68, 72, 74, 78, 83, 86, 93, 126, 127.

Habitat: In sand, in spinifex shrubland.

Size and Form: Shrub, up to 3 m tall, sometimes with drooping branches.

Leaves: Usually leafless, except on flowering branches. Narrow, rather like pine-needles, 3 – 5 mm long, terminating in a distinct point. Pale yellow-green.

Flowers: Arranged in spikes up to 16 mm long. Yellow-green, occurring in notches. Petal lobes 1 mm long.

Fruit: Oval, 3 – 5 mm long, with longitudinal ribs. Borne on a fleshy pink-red edible stalk, 2 mm long.

Uses: Women and children collect great quantities of this fruit seasonally, by spreading skin cloaks beneath the bushes, which are shaken to dislodge the ripe fruit. It tastes slightly acidic, rather like a currant.

Comments: The fruit was used by early settlers to make jam.

CREEK FIG

Scientific Name:	*Ficus coronulata*
Aboriginal Name:	DJABAWING (Kununurra)
References:	58, 82, 86, 87, 102
Habitat:	Along watercourses in or near rainforest, usually near the coast.
Size and Form:	Shrub or tree, up to 12 m tall. With smooth bark and drooping branches but no aerial roots. Separate male and female plants.
Leaves:	Oblong or oval, 7 – 15 cm long, 1.5 – 6 cm wide, sometimes lobed. Rough on upper surface.
Flowers:	Inside the fruit. Male plants produce figs containing male flowers, female plants produce figs containing bisexual and female flowers.
Fruit:	Globular, 8 – 20 mm diameter, usually covered in dense hairs. Green-yellow and pleasantly flavoured when ripe.
Uses:	The green fruit is eaten during the wet season. The milky latex of young shoots is used in the healing of wounds.
Comments:	The flowers of *Ficus* spp. are produced inside the fruit. Insects must penetrate the fig to pollinate the flowers, which then develop into the structure that is eaten.

SWAMP FIG

Scientific Name:	*Ficus hispida* var. *hispida*
Aboriginal Name:	KWURNINYARRA (Fitzroy Crossing)
References:	58, 63, 113.
Habitat:	Swampy jungle.
Size and Form:	Tree, up to 15 m tall. Bark pale grey, roots slightly buttressed. Lower branches arch down to ground.
Leaves:	Oval or ovate, 22 cm long, 11.5 cm wide, with prominent veins. Dark-green, rough.
Flowers:	Inside the fruit. Male and female figs borne on separate plants.
Fruit:	Globular, up to 4 cm in diameter, black when ripe. Large in comparison with those of other bush figs. Borne in clusters on fruiting spurs.
Uses:	The figs are eaten.

LARGE-LEAF ROCK FIG

Scientific Name: *Ficus leucotricha*

Aboriginal Name: PINGALEE (Kalumburu)

References: 58, 63, 87, 102, 113

Habitat: Grows out of cracks in rock outcrops, hence the common name. Often with aerial roots.

Size and Form: Tree, up to 7 m tall, with rounded crown of dense foliage 5 – 6 m wide.

Leaves: Ovate, up to 12 cm long, 8 cm wide. Bright green upper surface, greyish underside. Covered with fine velvety hairs.

Flowers: Inside the fig 'fruit' structure, as in all *Ficus* spp.

Fruit: Globular. Yellow, becoming red. Borne at the ends of branches. Producing a milky sap.

Uses: The fruit is edible.

Comments: Cattle readily eat the foliage. There are several varieties recorded for the area.

SANDPAPER FIG

Scientific Name:	*Ficus opposita* var. *indecora*
Aboriginal Name:	MUTHI (Ramingining)
References:	12, 59, 92, 102, 107, 113
Habitat:	Dry jungle. Sometimes in open forest or on river banks.
Size and Form:	Small straggly tree, up to 8 m tall.
Leaves:	Oval to ovate, up to 10 cm long, 7 cm wide. Stiff, dark-green, covered with short spiny hairs.
Flowers:	Inside female 'fruit' structure. Separate male and female plants.
Fruit:	Globular, up to 10 mm in diameter, rough. Green; black when ripe. Borne on old growth, usually appearing after leaves shed.
Uses:	The fruit is eaten raw. Leaves are used to sand spears. Leaves are heated on hot stones and applied to reduce swelling. Infusions of bark are used to treat diarrhoea. In cases of ringworm, the leaves are used to roughen the skin, then a decoction of *Passiflora foetida* is applied.
Comments:	Another sandpaper fig, *Ficus scobina* (MANLARRA, Kununurra), grows amongst sandstone or granitic rocks. Its fruit is eaten during early wet.

ROCK FIG

Scientific Name: *Ficus platypoda*

Aboriginal Name: YILI (desert areas)

References: 12, 13, 16, 38, 47, 62, 74, 76, 126

Habitat: Widespread on rocky outcrops and in gullies on rocky hillsides, around water holes and occasionally on flat country.

Size and Form: Sprawling tree or shrub, 3 m tall, occasionally to 10 m. Often with several stems and many spreading branches.

Leaves: Oval with distinct point, 5 – 10 cm long, 1.5 – 5 cm wide. Dark-green, shiny, tough and thick. Arranged at ends of branches on broad stalks.

Flowers: Inside small fruits clustered towards ends of branches.

Fruit: Globular figs, 1 – 2 cm in diameter. Dark red to black, covered with small yellow dots when ripe. Usually in pairs between base of leaf and branchlet.

Uses: The fruit ripens during the wet season and is available in March and April. It is eaten raw, dried, or ground to a paste with water or a little bush honey. Can be used for making jam.

Comments: These trees hold a special place in Aboriginal lore. According to traditional belief, the fruit of this sacred tree should not be stewed or cooked.

CLUSTER FIG

Scientific Name: *Ficus racemosa*

Aboriginal Name: ONGURR (Kalumburu)

References: 2, 12, 32, 58, 87, 92, 102, 113, 126

Habitat: River banks, coastal monsoon forest, edges of coastal plains.

Size and Form: Spreading tree, up to 30 m tall. Trunk yellow-green to grey, smooth bark.

Leaves: Oval, tapered at both ends, 8 – 15 cm long, 3 – 7 cm wide. Dark green, papery in texture. Stalks 2 – 5 cm long. Borne singly at different heights on axis.

Flowers: Inside the female fruiting structure. Separate male and female fruits.

Fruit: Urn-shaped, fleshy, up to 5 cm in diameter. Yellow, becoming reddish when ripe. Borne on the trunk, stems and branches rather than at the extremities of the branches as in other figs.

Uses: The fruit is edible. Cattle eat the leaves of smaller plants. Dugout canoes are sometimes made from the trunks of these trees.

Comments: When the bark is wounded it discharges a white milky sap.

FLAGELLARIA

Scientific Name: *Flagellaria indica*

Aboriginal Name: YUROL (Kimberley)

References: 1, 54, 102.

Habitat: Within rainforest or along the margins, and along streams.

Size and Form: Climbing perennial, up to 10 m tall. Woody stems, 1 – 3 cm thick.

Leaves: Narrow, tapered at both ends, 10 – 30 cm long. Bright green.

Flowers: Small, white, borne in open much-branched clusters at the ends of stems.

Fruit: Round berries, red when ripe, small and fleshy.

Uses: The young shoots are edible after boiling. The fruit is edible. The stems are used for split cane work (fish traps, baskets, armbands etc.) and for making handles for baskets, brushes and other artefacts. Some people use this plant medicinally to provide relief from a number of ailments. Infusions of the growing tips and leaves are taken for chest complaints, throat ailments and toothache.

WHITE RAISIN

Scientific Name:	*Flueggia virosa*
Aboriginal Name:	GUWAL (Broome)
References:	2, 63, 87, 102, 113
Habitat:	Widespread on sands near water.
Size and Form:	Large shrub or small tree, up to 4 m tall.
Leaves:	Oval to ovate, 9 cm long, 5 cm wide. Arranged alternately on opposite sides of the stem in the same plane.
Flowers:	White, unisexual, scented, borne in dense clusters.
Fruit:	Fleshy succulent globular capsules, 5 mm in diameter. White when mature.
Uses:	An infusion of the young leaves is drunk for internal pains. The liquid is applied externally for open sores and leprosy.
Comments:	Aerial parts of the plant contain several alkaloids. Leprosy is now contained in Australia and the last leprosarium, near Derby, has closed.

WILD GARDENIA

Scientific Name: *Gardenia megasperma*

Aboriginal Name: MANGONG (Kununurra)

References: 58, 92, 102, 113

Habitat: Amongst sandstone rocks. Open woodland near creeks.

Size and Form: Shrub or small tree, 4 – 5 m tall, with thick branches. Bark mottled yellow and grey.

Leaves: Oval, 8 – 14 cm long, 6 – 11 cm wide. Velvety when young, becoming less so as they mature.

Flowers: Short, trumpet-shaped, with about seven 'petals', 3 – 5 cm in diameter. White, sweetly scented. Borne singly at ends of branches.

Fruit: Oblong, 5 – 7 cm long. Hard, containing many seeds embedded in floury pulp.

Uses: The red gum is used as a glue. A decoction of the whole fruit is used to treat toothache. The flesh of the fruit is edible. The hard white wood is sometimes used for artefacts.

ROCK GREVILLEA

Scientific Name:	*Grevillea heliosperma*
Aboriginal Name:	(Not collected)
References:	1, 32, 65, 113
Habitat:	Rocky hillsides and outcrops.
Size and Form:	Straggly, tree-like shrub, up to 7 m tall, usually smaller.
Leaves:	Divided into six leaflets, each leaflet distinct and resembling a separate leaf. Leaf blade up to 20 cm long, stem 7 cm long. Grey-green, margins smooth.
Flowers:	Bright red, ageing to pink, 12 mm long. Borne in branching spikes up to 15 cm long.
Fruit:	Round hard woody capsules, with persistent stigma protruding from end. Opening in September – October to reveal two heart-shaped seeds.
Uses:	In winter, the leaves and bark are heated to produce a bright red liquid, which is used to bathe sores or for general illness. The seeds and gum are eaten.
Comments:	Near Kununurra, this species could be confused with a similar species, *G. decurrens*, which has greyish leaves and less crowded cream to pinkish flower spikes.

CAUSTIC BUSH

Scientific Name: *Grevillea pyramidalis*

Aboriginal Name: TJIINNGU (Roebourne)

References: 32, 63, 87, 113

Habitat: Rocky hills, along the slopes of tablelands and on sandplains.

Size and Form: Shrub or tree, up to 4 m tall. Bark corky and fissured. Branchlets and leaves sparsely hairy.

Leaves: Long and narrow, 10 – 35 cm long. Deeply divided into pairs. Several pairs arranged into a spray on a single 'stem'. Silvery to blue-green. Hanging downwards.

Flowers: Small, white or cream. Clustered into pyramid shapes, up to 30 cm long, at the ends of branchlets. Flowering late winter and spring.

Fruit: Flat oval woody pea-like pods, 2 – 2.5 cm long. Covered with brown-black sticky varnish-like gum that causes the skin to blister.

Uses: The greenish inner bark is mashed with water until it is white, and the infusion rubbed around the breast to induce lactation. A solution of peeled inner bark soaked in water is dabbed on sores.

Comments: The caustic sap that covers the seed cases is used to make skin markings.

BEEFWOOD

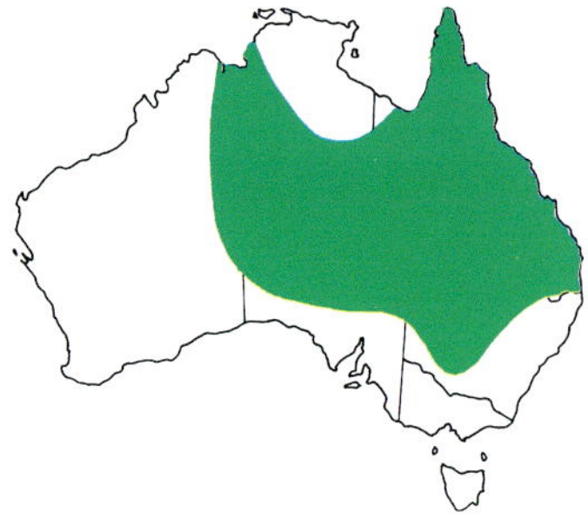

Scientific Name:	*Grevillea striata*
Aboriginal Name:	ILTILPA (Wiluna)
References:	32, 75, 113
Habitat:	Alluvial sand, loam or gravel. On plains, by creeks, rarely on hills.
Size and Form:	Shrub or tree, up to 9 m tall. Trunk erect and straight. Bark fissured, dark-grey.
Leaves:	Narrow, 10 – 30 cm long, flat, with prominent veins on underside.
Flowers:	Small, off-white, arranged in erect branched clusters 5 – 8 cm long. Flowering early summer.
Fruit:	Broad, oval, woody, like pea pods, 1 – 5 cm long.
Uses:	Dried sap is collected from damaged areas of bark. It is ground to a powder and sprinkled onto sores, burns and cuts as a drying agent. The sap can also be used as a cement in the making of artefacts.
Comments:	Slabs of wood are often split from the trunk and used to make shields and woomeras. Throughout the interior of Western Australia, numbers of these trees have been blazed by early European settlers to mark survey points and camps.

HOLLY-LEAFED GREVILLEA

Scientific Name:	*Grevillea wickhamii*
Aboriginal Name:	YINNGINGA (Balgo area)
References:	28, 85.
Habitat:	Sandy or gravelly plains. Dunes and rocky hills.
Size and Form:	Erect shrub, up to 5 m tall. Bark white and smooth, becoming dark and rough with age.
Leaves:	Seven-lobed with pointed spines, 3 – 8 cm long. Wavy margins. Greyish to dull olive-green.
Flowers:	Red-orange spidery flowers, 4 – 5 mm long, with curved yellow tips. Arranged in showy bunches, 3 – 5 cm long. Scattered over whole shrub.
Fruit:	Leathery capsules, up to 5 cm long. Seeds 1.5 – 2 cm long, shed as soon as capsules mature.
Uses:	Gum exuded from fissures in the trunk is eaten by Aborigines in the Balgo area.
Comments:	Freshly opened flower bunches contain sweet nectar, which can be sucked straight from the flowers or shaken into a container.

DYSENTERY BUSH

Scientific Name: *Grewia retusifolia*

Aboriginal Name: MALAGANYE (Derby area)

References: 98 (related species)

Habitat: Open forest and sandy ground near watercourses in coastal areas.

Size and Form: Straggling shrub, up to 1 m tall, with hairy non-branching stems.

Leaves: Oval, up to 14.5 cm long. Rough, with prominent veins, underside greyish and densely hairy.

Flowers: About 9 mm in diameter, with about 20 protruding stamens and five long white 'petals'.

Fruit: Small brown two-lobed berries. Thin papery skin and thin flesh covering seeds.

Uses: The fruit is eaten raw to cure diarrhoea and dysentery, which can also be treated by chewing the leaves or drinking an infusion of the mashed roots. The infusion is also used as an eyewash.

Comments: The fruits are called 'jelly-boys'. This species has not been clearly distinguished from *Grewia polygama* in the Kimberley.

STINKWOOD

Scientific Name: *Gyrocarpus americanus*

Aboriginal Name: DJALALANG (Kununurra)

References: 12, 58, 93

Habitat: Alluvial plains in areas of medium rainfall. Often associated with limestone.

Size and Form: Small to medium-sized tree, up to 12 m tall. Stout smooth trunk, deciduous leaves crowded at the ends of branches.

Leaves: Ovate to circular, green, often 25 – 30 cm wide. On young trees leaves have three deep lobes, on older trees they are undivided and heart-shaped.

Flowers: Cream, 2 – 3 mm in diameter. Borne in dense clusters in axils of upper leaves.

Fruit: Two large wings, 3 – 8 cm long, may carry the seed in the wind.

Uses: An infusion made from the roots and young shoots is rubbed onto cuts (but not open wounds) and onto rheumatic joints. Powdered charcoal made from the wood is used to heal cuts. The wood is light, and is used to produce a range of artefacts, especially deep bowls or 'COOLAMONS'.

Comments: There are several subspecies of this tree.

BLOODROOT

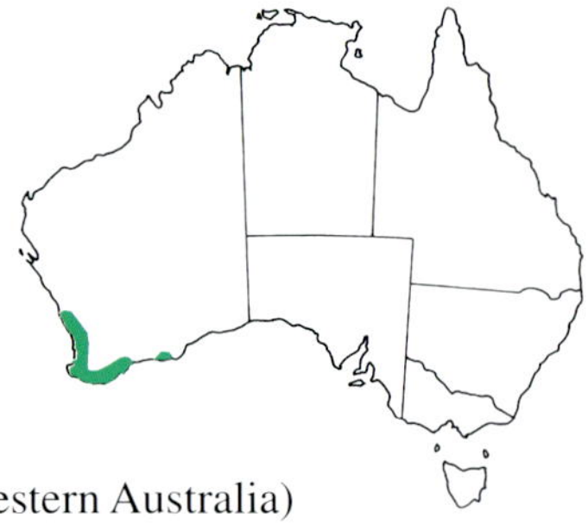

Scientific Name: *Haemodorum spicatum*

Aboriginal Name: BOHN, KOOLUNG, MEEN (south-west Western Australia)

References: 18, 68, 72, 83, 86, 93

Habitat: Kwongan and woodland. Widespread and common, often forming thick stands, especially near swamps, creeks or around granite rocks.

Size and Form: Herbaceous perennial. Foliage up to 1.5 m tall, arising from a blood-red bulb growing at or just below soil level.

Leaves: At base of plant, 3 – 5 narrow linear cylindrical leaves, 20 – 80 cm long. Small leaves borne singly at different heights along the stems.

Flowers: Scattered along erect spikelet, up to 2 m tall. Dark purple-black.

Fruit: Small red capsules held on extremities of dried flower stems. Three slits open in capsule, releasing flat ovate black seeds. Old seed stalks and capsules are black.

Uses: Used to cure dysentery. The roots are eaten raw or roasted. They are sweet juicy and hot, somewhat resembling a very mild onion in flavour, but with considerable 'bite'. The bulb-like leaf bases are roasted, pounded together with clay from termite nests and eaten.

Comments: Several *Haemodorum* species were eaten in the south-west of Western Australia. It is now difficult to associate the relevant Aboriginal names with the different species.

YELLOW HAKEA

Scientific Name: *Hakea arborescens*

Aboriginal Name: DERINDING (Kununurra)

References: 58

Habitat: Open forest, in seasonally swampy soil.

Size and Form: Small woody tree, up to 7 m tall. Bark rough corky, almost black.

Leaves: Strap-like, tapered gently to the base, may be curved near apex.

Flowers: White or pink-white, 4 mm wide, in spidery clusters in leaf axils.

Fruit: Woody, 5 cm long, 3 cm wide, splitting to reveal two seeds. Seeds triangular with sharp point. Large oval wing opposite point.

Uses: The seeds are eaten by children. Leaves and green pods are boiled in water to produce a red-brown liquid, which is used to bathe scabies and sores. It is also used as a rub for muscular aches and pains.

Comments: The wood has been used to make boomerangs.

CORKWOOD TREE

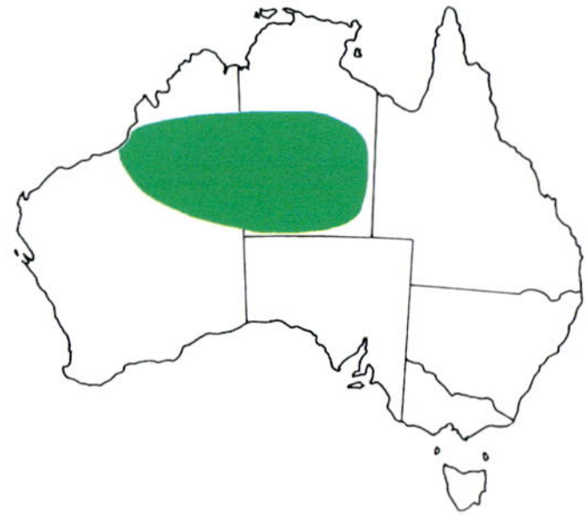

Scientific Name:	*Hakea macrocarpa*
Aboriginal Name:	DYARIDANY (Broome)
References:	32, 91, 98, 102
Habitat:	Sandy coastal plains in arid regions.
Size and Form:	Small tree, 3 – 8 m tall. Stout trunk, rugged fissured bark.
Leaves:	Narrow, tapered at both ends, 15 – 20 cm long. Thick, hairy on both sides, borne singly at different heights on the axis.
Flowers:	Green-yellow, in clusters 7 – 15 cm long.
Fruit:	Egg-shaped, tapered towards tip, 3.5 – 4 cm long.
Uses:	Charcoal made from the wood is applied to open sores and cuts. The bark is burnt to a black powder and applied to sore lips. The wood is used to make artefacts.

CORKBARK

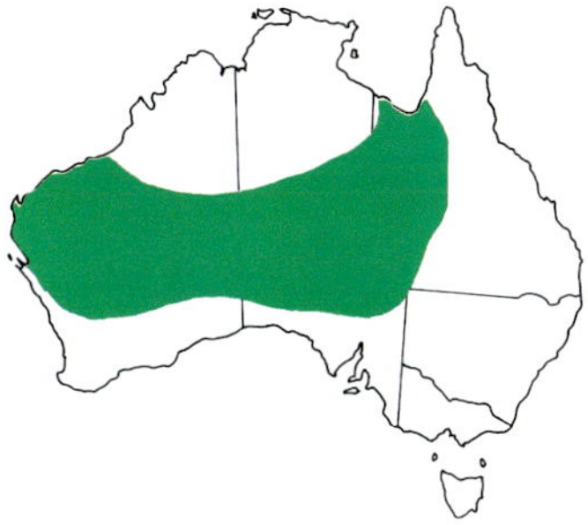

Scientific Name:	*Hakea suberea*
Aboriginal Name:	PARNTAL (Wiluna)
References:	32, 75, 82, 113
Habitat:	Deep sandy loam, often on the flood plains of watercourses.
Size and Form:	Tree, up to 7 m tall. Trunk and branches covered with deeply fissured bark. Branches twisted or contorted.
Leaves:	Needle-like with sharp points, up to 30 cm long. Slightly woolly when young but becoming smooth with age.
Flowers:	Deep cream-yellow, hairy, 1 – 1.2 cm long. In cylindrical clusters, 8 – 16 cm long.
Fruit:	Woody capsules, egg-shaped and somewhat flattened, 3.5 cm long, 2.5 cm wide. Persisting for several years on tree.
Uses:	On opening, the flowers are rich in sweet nectar, which is dissolved in water to produce a sweet drink.
Comments:	The flowers of many hakeas (and grevilleas) produce nectar in profusion. This is collected avidly, either by sucking directly from the flowers or by dissolving it from the flowers in dishes or billycans of water. The latter is recommended, as the flower parts taste bitter.

BUNCH SPEAR GRASS

Scientific Name: *Heteropogon contortus*

Aboriginal Name: NGALARA (Broome)

References: 12, 47, 102

Habitat: A common roadside weed. Also on red lateritic sands and loams over deeper clays.

Size and Form: Grass, forming open tussocks. Flower stems fine, branching, up to 1 m tall.

Leaves: Bright green, wiry, 6 mm wide.

Flowers: Several small 'flowers' on each branch of flower stem.

Fruit: Long, narrow, pointed, with a long twisted dark-brown appendage, up to 10 cm long.

Uses: The leaves are macerated and soaked or boiled and the hot liquid used to relieve cough symptoms. The whole plant can be burnt as a mosquito repellent.

Comments: As with most grasses, the flower-heads, which produce the seed, are not recognised as flowers. The long appendages on the seeds often twist together into a matted clump held in bunches above the plants.

YELLOW HIBISCUS

Scientific Name:	*Hibiscus panduriformis*
Aboriginal Name:	(Not collected)
References:	12, 13, 47, 51, 87, 102, 113
Habitat:	Widespread in a variety of habitats, particularly river banks and beds of intermittent streams.
Size and Form:	Shrub, 1 – 2.5 m tall, with soft 'felty' greyish hairs.
Leaves:	Rounded, 4 – 9 cm in diameter, with heart-shaped base. Crinkly margins, pale underside.
Flowers:	Bell shaped, 10 cm in diameter. Striking, bright-yellow with crimson centre. Flowering in late wet season.
Fruit:	Capsules, 1 – 2 cm long, covered with dense straw-coloured hairs. Seeds sparsely covered with short soft hairs.
Uses:	The centres of the flowers can be picked out early in the morning and eaten.
Comments:	Some references to string-making in the Pilbara region suggest that the inner bark is suitable for this purpose.

A COWVINE

Scientific Name:	*Ipomoea colobra*
Aboriginal Name:	INTAL (Ashburton area)
References:	83, 93
Habitat:	Red sandy or clay soils. Amongst mulga thickets.
Size and Form:	Robust deciduous climber, up to 4 m tall. Root tubers black, 2 – 25 cm long, 2 – 4 cm wide.
Leaves:	Heart-shaped, 5 – 8 cm long, 20 – 40 cm wide. Growing singly and on alternate sides of the creeping stems.
Flowers:	Petals pink-white. Five narrow oval sepals. Borne singly or in groups of 2 – 6, on short stalks at the angle between leaf and stem.
Fruit:	Brittle egg-shaped capsules with a translucent covering, containing 4 – 10 seeds. Seeds egg-shaped or triangular, 3 mm in diameter, sparsely covered with short soft hairs.
Uses:	The tuber, which resembles and tastes like a sweet potato, is edible if baked in hot ashes and sand.
Comments:	The application of this name to the species in Queensland is erroneous.

BUSH SWEET POTATO

Scientific Name:	*Ipomoea costata*
Aboriginal Name:	KARNTI (Balgo area)
References:	24, 43, 62, 69, 93.
Habitat:	Sandy soils, especially near creek beds and watercourses.
Size and Form:	Climber, up to 3 m tall. Large sweet-potato-like tuber, buried 20 – 30 cm below the surface.
Leaves:	Heart-shaped, 4 – 7 cm long.
Flowers:	Funnel-shaped, purple-pink, 4 – 5 cm in diameter.
Fruit:	Capsules with transparent walls containing four or five black seeds with white fluffy covering.
Uses:	The tuber is roasted in hot coals and eaten.
Comments:	Tuber-bearing roots have a distinctive 'raw potato' smell. They are located by searching for cracks radiating from the base of the plant; indicating disturbance below the ground.

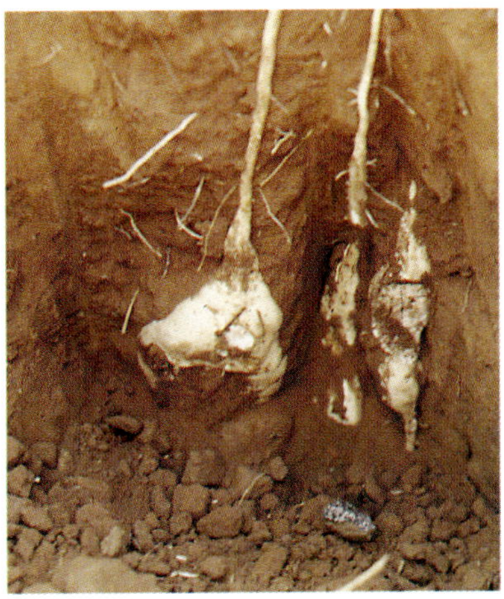

PURPLE BEACH CONVOLVULUS

Scientific Name: *Ipomoea pes-caprae braziliensis*

Aboriginal Name: YINGA (Kalumburu)

References: 12, 13, 24, 60, 65

Habitat: Along beaches, on coastal cliffs.

Size and Form: Perennial herb with creeping stems running over sand, often forming a mat. Stems 10 m or longer.

Leaves: Oval, sometimes broadened towards apex, notched or two-lobed, 5 – 8 cm long. Borne singly at different heights on the axis.

Flowers: Pink, trumpet-shaped, broad, five-sided, 4 – 12 cm in diameter. Borne on long stalks.

Fruit: Globular two-celled capsules, containing 2 – 4 woolly seeds.

Uses: A leaf decoction (or heated leaves) is applied to sores, scabies, bites and stings.

Comments: The swollen roots of several *Ipomoea* species are edible, but tuberous roots suitable for testing have not been found on this species in Western Australia.

ROCK ISOTOME

Scientific Name: *Isotoma petraea*

Aboriginal Name: TUNDIWARI (central desert)

References: 51, 60, 69, 74, 78, 82, 83, 93, 127

Habitat: Rock crevices and caves. Often on granite rocks in the dry interior.

Size and Form: Bushy perennial or annual herb arising from a woody taproot, 15 – 30 cm tall.

Leaves: Broad oblong, 3 – 8 cm long. Margins bearing irregular sharp teeth. Borne singly at different heights on axis.

Flowers: Pale blue, white or pink. Five petals, 2 – 3 cm diameter.

Fruit: Egg-shaped capsules, 1 – 2 cm long.

Uses: The plant is dried over the fire and pounded into a powder with ash from the bark of *Acacia aneura*. Small amounts are taken to relieve cold symptoms. A substitute for chewing tobacco. Referred to as 'little poison'.

Comments: The sap is very bitter, and irritating to the eyes.

PRICKLY LETTUCE

Scientific Name: *Lactuca serriola*

Aboriginal Name: (Not collected)

References: 12, 13, 67, 83

Habitat: An introduced species, now a widespread weed on roadsides and wastelands.

Size and Form: Usually a single-stemmed herb, 1 – 2 m tall, arising from a flat rosette of leaves at ground level.

Leaves: Broadly oblong, up to 15 cm long, with deep lobes. Prickly margins, bristles on underside of main vein.

Flowers: Pale-yellow, daisy-like, 1 cm in diameter. Borne at the ends of stems.

Fruit: Seeds stand erect on small 'thistle-head' structures.

Uses: The young spring leaves are an excellent vegetable. Older leaves are very bitter. The roots are slender, woody and scarcely palatable.

Comments: The spiny stems ooze milky sap when cut.

COGOLA

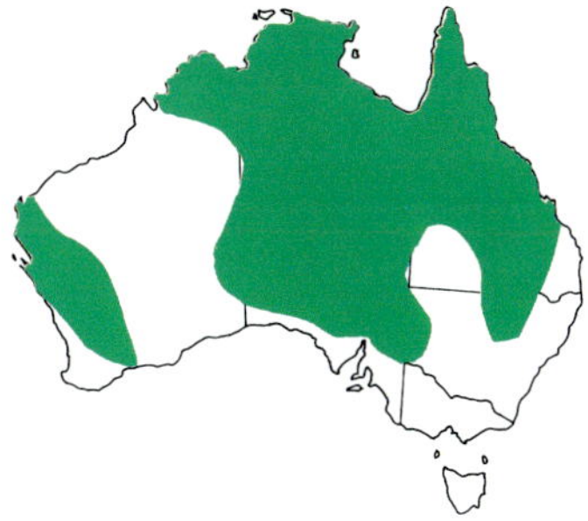

Scientific Name: *Leichardtia australis*

Aboriginal Name: KALKULA (Kalgoorlie)

References: 13, 16, 62, 72, 76, 78, 82, 93, 127.

Habitat: Near watercourses or rocky outcrops.

Size and Form: Loose climber, up to 2 m tall. Stems, branches and stalks covered with a greyish layer of very short closely woven hairs, giving it a silky appearance.

Leaves: Narrow, oblong, coming to a point, 4 – 10 cm long. Covered in greyish hairs when young.

Flowers: Flower heads composed of 5 – 10 green-yellow flowers. Arising from leaf axils.

Fruit: Seeds 2 – 4 cm long, with a long 'cone' of hairs.

Uses: The young fruit stalks and leaves have an agreeable sweetish taste. The fruits are edible when young, before they become dry and mature. Nectar is collected from the flowers and eaten.

Comments: Previously known as *Marsdenia australis*. The Aboriginal name for this plant gave the town of Kalgoorlie its name.

BEARDED HEATH

Scientific Name: *Leucopogon parviflorus*

Aboriginal Name: (Not collected)

References: 18, 25, 68, 86

Habitat: Coastal areas, in dune swales (the low parts between dune crests).

Size and Form: Shrub, up to 2.5 m tall. Usually much-branched and twiggy.

Leaves: Broad, tapering towards tip, 8 – 20 mm long, rigid.

Flowers: Tiny, tubular, 4 mm long, white. Characteristic white woolly mass of hairs on petal lobes.

Fruit: Rounded, 5 mm long, yellow-green to translucent.

Uses: The fruit is reported to be edible.

Comments: Many small berries and fruits are gathered by spreading a skin cloak beneath a plant and beating or shaking it to dislodge the ripe fruits, which collect on the cloak.

MILLSTREAM FAN-PALM

Scientific Name:	*Livistona alfredii*
Aboriginal Name:	(Not collected)
References:	52, 58
Habitat:	Semi-tropical regions, around permanent water.
Size and Form:	Fan palm, 6 m tall. Typical palm, with large leaves at top of stout trunk.
Leaves:	Similar to an unfurled circular fan, up to 1 m wide. Thick stalks with spiny margins.
Flowers:	Cream-white, 6 mm wide, borne in drooping clusters.
Fruit:	Globular, hard, reddish brown, 3 cm in diameter.
Uses:	The centre part or 'heart' of the palm can be cut out and eaten 'like cabbage', but this is not recommended as the practice destroys the palm.
Comments:	This species has a restricted distribution, and is found only at Millstream on the Fortescue River, Western Australia.

MITCHELL PLATEAU PALM

Scientific Name:	*Livistona eastonii*
Aboriginal Name:	TAANA (Kalumburu)
References:	87
Habitat:	Lateritic soils, often in dense stands. Also found as a co-dominant or dominant species in the understorey of eucalyptus woodlands.
Size and Form:	Fan palm, up to 12 m tall. Highly decorative persistent leaf bases give trunk a 'reptile skin' effect.
Leaves:	Fan-shaped, broadly circular, 1 m wide. On long stem.
Flowers:	Cream-white, in pendulous axillary clusters of several thousand flowers.
Fruit:	Dark, spherical, 6 mm in diameter.
Uses:	The leaves and lower stem of the growing bud or heart are chewed for their sweet taste. Palm 'heart' can be extracted from the growing point and has a delicious nutty flavour, but this is not recommended as it will kill the adult plant.
Comments:	Slow-growing and difficult to transplant past the seed-leaf stage.

SAND PALM, FAN PALM

Scientific Name: *Livistona humilis*

Aboriginal Name: WURRDJARA (Milingimbi)

References: 2, 52, 100, 102

Habitat: Very common in open forest, frequently growing either in scattered colonies or pure stands.

Size and Form: Single-stemmed dwarf fan-leaved palm. Leaves spreading, in a small sparse crown. Trunk 1 – 3 m tall, 8 – 12 cm thick, dark brown to black.

Leaves: Fan-shaped, 1 – 1.3 m long, bright green. Stalks 60 – 90 cm long, 3 cm wide, tapering to 1.5 cm wide at top.

Flowers: Shallow three-petalled cups, 2 – 4 mm in diameter, pale yellow. Almost stalkless, borne in dense pendulous clusters.

Fruit: Oval, 1 – 1.5 cm long. Black.

Uses: The inner part of the stem is crushed, soaked overnight, then boiled to make a decoction, which is drunk for coughs, colds and to clear the chest. The palm heart, which tastes like coconut, can also be extracted, though this is not recommended as it will kill the plant. Fibre for string-making is extracted from the dried leaves.

A MISTLETOE

Scientific Name:	*Lysiana exocarpi*
Aboriginal Name:	NGANYTJA (central desert)
References:	51, 123
Habitat:	Branch parasite found on shrubs and trees in most habitats. Hosts include *Acacia* spp., *Eremophila* spp. and *Senna* spp.
Size and Form:	Clumps hanging in host plants, up to 70 cm long, 30 cm wide.
Leaves:	Narrow, linear to oblong, 5 – 15 cm long.
Flowers:	Tubular, 25 – 50 mm long. Red, rarely yellow.
Fruit:	Soft, egg-shaped, berry-like, 6 – 10 mm long. Red or black, borne in clusters.
Uses:	The fruit is edible.
Comments:	There are several subspecies of this plant. It is not considered a major food plant, but the fruit is knocked down from trees and eaten by children.

MULGA MISTLETOE

Scientific Name: *Lysiana murrayi*

Aboriginal Name: PARKAPARKA (central desert)

References: 28, 38, 51, 123

Habitat: Branch parasite on acacia shrubs and other trees.

Size and Form: Grows out from the host tree as clusters of branches, which look as though they have been grafted on. Easily distinguished from the host by the different-coloured foliage.

Leaves: Linear or narrow, pointed tip, curly, somewhat fleshy, 2.5 – 6 cm long. Pale green.

Flowers: Tubular, 1 – 3.2 cm long. White, yellow or pink. Usually solitary but sometimes in pairs.

Fruit: Small fleshy globular berries, usually pink or red when ripe.

Uses: The fruit can be eaten.

Comments: An uncommon but widely distributed species.

BAUHINIA

Scientific Name: *Lysiphyllum cunninghamii*

Aboriginal Name: WARIMPO (Derby)

References: 1, 87, 102

Habitat: Pasture lands with black soil, water holes in clay, river banks and sands with clay at depth.

Size and Form: Tree, up to 10 m tall. Short trunk and irregular branching. Bark dark grey-black, hard and flaking.

Leaves: Butterfly-shaped, with two lobes, folded down the middle, 2 – 4 cm wide.

Flowers: Short, tubular structures with protruding stamens. Pink-red. Solitary, clustered on short stems.

Fruit: Pods, 8 – 12 cm long, 3 – 6 cm wide. Bright red when young, brown when mature.

Uses: The leaves and pods are eaten by stock. Nectar is extracted from the flowers and turned into a fermented drink. A 'wash' is made from a decoction of bark.

ZAMIA

Scientific Name: *Macrozamia riedlei*

Aboriginal Name: BOOYO (south-west Western Australia)

References: 18, 36, 47, 68, 83, 86, 102

Habitat: Sandy and lateritic soils in a wide variety of habitats, from kwongan to tall forests.

Size and Form: Palm-like, up to 3 m tall. Trunk cylindrical, or egg-shaped in shorter specimens. Some populations produce trunks 1.5 – 1.8 m tall.

Leaves: Palm-like, 10 – 30 per plant. Grey-green with irregular striations, 1 – 3 m long.

Flowers: Non-flowering plants, which reproduce using separate male and female cones. Wind carries the male pollen to the female cones.

Fruit: Male plants bear long narrow cones with microspores on the scale-leaves. Female plants have 1 – 3 terminal cones, which have an outward appearance like that of a large green pineapple. Seeds within the cone are red, roughly egg-shaped, 3 cm long, 1 cm wide.

Uses: The fruit is eaten by humans, but the seeds are toxic. The red outer covering is eaten by animals such as emus, and was also treated to render it non-toxic to humans.

Comments: The whole plant contains dangerous toxins, which must be leached out by a long and rather complex process before eating.

NARDOO

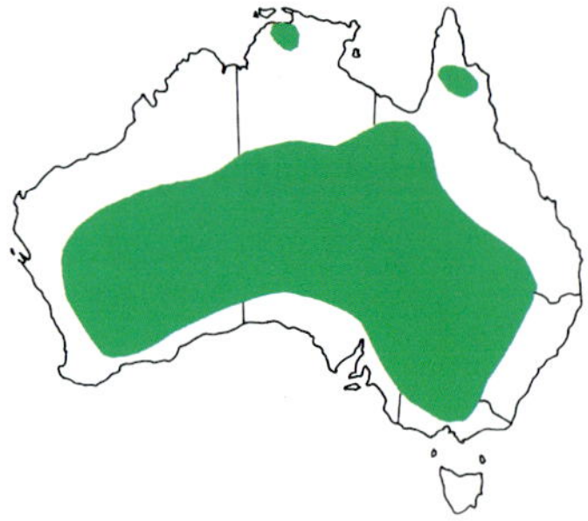

Scientific Name: *Marsilea drummondii*

Aboriginal Name: NARDOO (central Australia)

References: 12, 78, 83, 126, 127

Habitat: Seasonally wet areas.

Size and Form: Spreading water-loving plant. Leaves arranged at ends of stems 2 – 30 cm long. Stems branching from a creeping orange-brown hairy rhizome.

Leaves: Clover-like, divided into four leaflets, each leaflet triangular or tapering towards base, 30 cm long. Varying from almost hairless to densely hairy. Margins smooth or deeply scalloped.

Flowers: (None)

Fruit: Spore-bearing bodies (sporocarps) 4 – 9 mm long, ribbed, on stalks. Becoming hard and seed-like when mature, enclosing numerous spores.

Uses: The sporocarps are ground into flour and mixed with water to form an edible paste.

Comments: The rhizome may lie dormant for years between wet periods. Nardoo flour sustained some European explorers of Australia, but does not contain sufficient nutrition for prolonged unsupplemented consumption. Sporocarps of the short-fruited nardoo (*Marsilea hirsuta*) were probably also eaten.

WATTLE-FLOWERED PAPERBARK

Scientific Name: *Melaleuca acacioides*

Aboriginal Name: (Not collected)

References: 51, 65, 92, 102

Habitat: On or near river flats that are subject to saltwater flooding during very high tides in the wet season.

Size and Form: Upright straggly shrub or small tree, up to 2.4 m tall. Bark grey, thin, papery.

Leaves: Broad, tapering towards tip, terminating in a sharp thorn, 25 cm long, 7 cm wide. Green. Crushed leaves have a pungent aroma.

Flowers: White, with numerous stamens. Arranged in spikes, which are carried in dense ball-like clusters.

Fruit: Cup-shaped, up to 5 mm in diameter. Adhering to stem in rows.

Uses: An infusion of the leaves is drunk to relieve a sore belly. The aroma from the crushed leaves is inhaled for colds. Leaves also heated on the fire and rubbed on the forehead for headaches. Dry root sections are porous, and are used as friction firesticks; both drill and foot-piece.

SILVER CAJUPUT, RIVER PAPERBARK

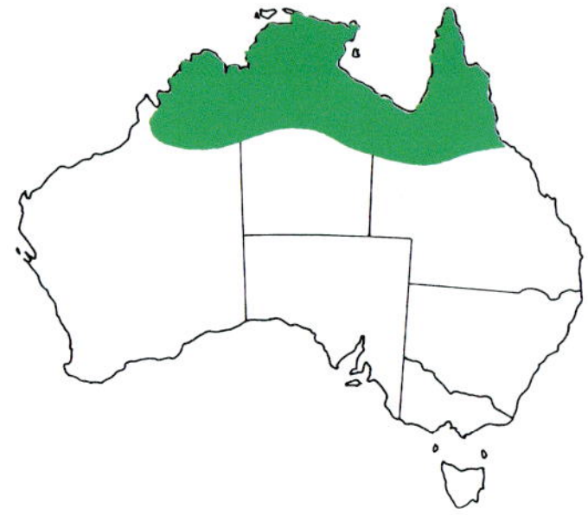

Scientific Name:	*Melaleuca argentea*
Aboriginal Name:	BANDARAN (Kalumburu)
References:	24, 45, 58, 76, 87, 102
Habitat:	Common on sandy river banks where the roots are in water at least for part of the year.
Size and Form:	Tree, 5 – 25 m tall. Silvery crown. Bark papery, in layers, smaller branches often drooping.
Leaves:	Narrow, 8 – 12 cm long. Silvery grey to pale-green.
Flowers:	Greenish, turning cream. Held in dense bottlebrush-like spikes, up to 8 cm long.
Fruit:	Cup-shaped, 3 – 4 mm wide, in clusters.
Uses:	The bark is used to wrap food for cooking, as a mat, as wrapping, and to cover wet-season shelters.
Comments:	Pieces of bark big enough to be used as 'blankets' can be stripped from large trees. Flying foxes like to 'camp' in the dense foliage of this species.

WHITE PAPERBARK

Scientific Name: *Melaleuca leucadendra*

Aboriginal Name: THALGO (Roebourne)

References: 51, 60, 65, 87, 113

Habitat: River banks and watercourses, coastal and freshwater swamps. Sometimes in deep sandy soil on beaches.

Size and Form: Tall upright tree, up to 30 m tall. Crown flattened or rounded, drooping twigs and thin papery grey-white bark.

Leaves: Narrow oval to narrow ovate, tapered towards tip, 16 mm long.

Flowers: White, heavily scented, 12 – 18 cm long. Carried stiffly at an angle to drooping stem, in groups of three, at 5 – 10 mm intervals along stem.

Fruit: Globular capsules, 5 mm in diameter, adhering to stem.

Uses: The bark is used to make waterproof shelters. Thinner pieces are used to wrap items. Children suck nectar from the flowers. The leaves are crushed, soaked and the decoction drunk for coughs and colds.

Comments: The trunks are reputed to be used to make canoes.

YELLOW-FLOWERING PAPERBARK

Scientific Name:	*Melaleuca nervosa*
Aboriginal Name:	PARKALI (Wave Hill)
References:	45, 51, 102
Habitat:	River banks.
Size and Form:	Small tree, up to 10 m tall. Young growth covered with soft hair.
Leaves:	Narrow, tapering to a point, 5 – 8 cm long, 1 – 3 cm wide. Dark green. Borne singly at different heights along branches.
Flowers:	Spikes, up to 9 cm long. Usually yellow-green, sometimes red.
Fruit:	Cup-shaped, 3 – 4 mm long. Borne in clusters up to 8 cm long along branches.
Uses:	The young leaves are crushed, boiled and used to bathe a patient for sores or general sickness. In the Northern Territory, the leaves are crushed with the bark of *Sesbania formosa* and roots of *Pandanus* spp. to make a 'wash'.

BROAD-LEAVED PAPERBARK

Scientific Name: *Melaleuca viridiflora*

Aboriginal Name: GNALIE (Kalumburu)

References: 51, 65, 87, 92

Habitat: Open forest and freshwater swamps, often in association with *Banksia dentata*.

Size and Form: Low shrub when growing in open forest, tree when growing in swamps and freshwater billabongs.

Leaves: Oval or narrow oval, 18 cm long, 5.6 cm wide. Carried stiffly on stems.

Flowers: Sepals and petals hidden by bunches of white to green-white stamens, which can be 22 mm long. Borne in bottlebrush-like clusters.

Fruit: More or less globular capsules, up to 7 mm in diameter. Borne in rows along the stems.

Uses: Women used the bark to make 'dress-barks' and waterproof shelters. A decoction made from the leaves is drunk for coughs and colds. The wood is sometimes used for making dugout canoes.

ORANGE SPIKE BERRY

Scientific Name:	*Mimusops elengi*
Aboriginal Name:	DJUNGUN (Broome)
References:	65, 68, 98
Habitat:	Sand dunes. Sandy banks of tidal rivers near mangroves and vine thickets.
Size and Form:	Small tree, up to 5 m tall. Umbrella-shaped.
Leaves:	Dark green, oval, shiny, 10 cm long, 4 cm wide.
Flowers:	Star-shaped, up to 16 mm in diameter. Petals and stamens white, forming a central spike. Calyx composed of thin sepals, giving a spiky appearance.
Fruit:	Round orange berries. Persistent calyx forms spiky cup.
Uses:	The fruit, which is rather dry, is eaten raw.
Comments:	This is a widespread species throughout the tropics of Africa, southern Asia, Australasia and the Pacific islands.

DESERT MOCK IVY

Scientific Name:	*Mukia maderaspatana*
Aboriginal Name:	WANTA-KAPI-KAPI (Balgo)
References:	12, 13, 47, 74, 76, 84, 85, 102, 127
Habitat:	Various widespread shrubby areas, particularly in thickly vegetated zones, but not in deep shade.
Size and Form:	Climbing herb with unbranched tendrils, attached to other plants.
Leaves:	Almost triangular, but varied in shape. With two broad lobes, 2 – 7 cm long. Almost stalkless, growing directly on trailing stems.
Flowers:	Yellow, up to 4 mm in diameter. Separate male and female flowers.
Fruit:	Globular, 8 – 12 mm diameter. Ripening to red. Containing 2 – 6 smooth pale-brown to dark-grey seeds with thickened margins. Inedible.
Uses:	The trailing stems are wound around the forehead as a headache cure.
Comments:	There are a number of species used in 'binding' or 'tying' medicine. It has not been tested whether any active chemical principles pass through the skin from the plant, but this type of traditional treatment remains popular.

PINK SWAMP LILY

Scientific Name:	*Murdannia graminea*
Aboriginal Name:	PANIYU (Kalumburu)
References:	2, 12, 24, 102
Habitat:	Found in association with *Cartonema parviflorum* in seasonally moist or wet areas.
Size and Form:	Perennial herb, up to 80 cm tall. Stem thin, sappy. Swollen tubers radiate from the central stem about 15 cm below ground surface.
Leaves:	Short, grass-like, up to 80 cm tall. Sometimes rough to the touch.
Flowers:	Three-petalled, 1 cm in diameter. Bright pink, mauve or bluish. Borne in sprays at tip of stem.
Fruit:	Oval capsules containing three cells, up to 12 mm long.
Uses:	The tubers are cooked in hot ashes and eaten.
Comments:	A variable species. See ref. 24 for another description of the edibility of this species. A related species, *M. nudiflora*, also occurs in the Kimberley region, possibly naturalised from Asia.

LEICHARDT TREE

Scientific Name: *Nauclea orientalis*

Aboriginal Name: MARR (Kalumburu)

References: 78, 87, 102

Habitat: Medium to high rainfall zones, along river banks and seepage areas.

Size and Form: Handsome riverbank tree, up to 25 m tall. Foliage dark green, timber dark yellow.

Leaves: Oval, 10 – 20 cm long, with distinct veins.

Flowers: Cream, arranged in spherical flower-heads, 3 cm diameter.

Fruit: Globular, 3 cm in diameter. Hard, brown, rough.

Uses: The fruit is edible but rather dry. An infusion of the bark has been used to induce vomiting for stomach disorders, and also as a fish poison.

Comments: The buds and new leaves are protected by large distinctive scales. A related species is used in traditional African medicines.

BUSH TOBACCO

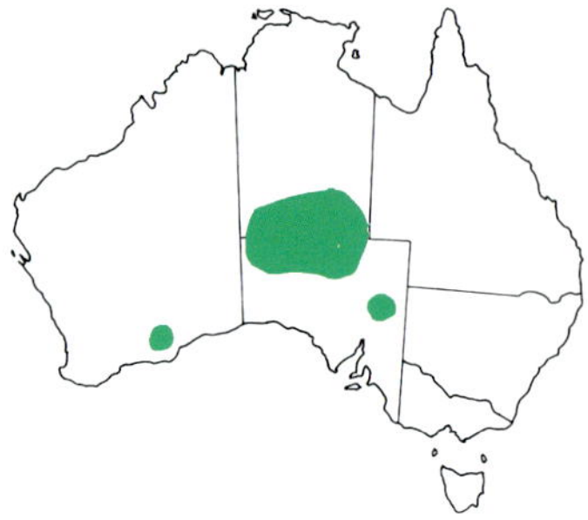

Scientific Name: *Nicotiana excelsior*

Aboriginal Name: PALANDU (Balgo)

References: 16, 17, 36, 51, 60, 107

Habitat: On rocky ground and around rock shelters in the ranges of central Australia.

Size and Form: Leafy herb, up to 1 m or taller. Taller parts arising from a rosette of hairless leaves at ground level.

Leaves: Dull green, up to 20 cm long. Smooth, with a few soft spines. Held on pendulous stalks. Borne singly at different heights on the axis.

Flowers: Whitish, tubular with trumpet-like tip, up to 6 cm long.

Fruit: Capsules 12 – 20 mm long. Seeds kidney-shaped, rounded or angled.

Uses: The narcotic leaves are mixed with ash from the bark of *Eucalyptus papuana* and chewed. A poultice of chewed leaves is sometimes applied to skin lesions.

SANDHILL TOBACCO

Scientific Name: *Nicotiana gossei*

Aboriginal Name: INGULBA, MINGULBA (central Australia)

References: 16, 73, 112

Habitat: Pockets of fertile, often sandy, soil in sheltered rocky areas.

Size and Form: Leafy herb, up to 2 m tall. Densely covered in woolly, glandular hairs.

Leaves: Broad with blunt tip, 25 cm long. Basal parts of leaf-blades extending downwards along stems. Soft, covered with soft but not sticky hairs. Borne singly at different heights on the axis.

Flowers: White. Hairy tube, 6 cm long, tipped by five broad blunt lobes.

Fruit: Capsules, up to 16 mm long, containing numerous tiny pitted seeds.

Uses: The leaves and stems are pounded with ash from leaves of *Acacia* spp., *Grevillea* spp. or the bark of eucalypts. The mixture is chewed for its narcotic effects.

WESTERN AUSTRALIAN CHRISTMAS TREE

Scientific Name: *Nuytsia floribunda*

Aboriginal Name: MUNGAH (south-west Western Australia)

References: 18, 47, 68, 72, 83, 86, 93, 102, 126

Habitat: Common in sandy soil.

Size and Form: Tree, up to 8 m tall, with suckering roots. Grey outer and pale-yellow inner bark.

Leaves: Scale-like on young plants. Linear and almost fleshy on older plants, 100 mm long, 8 mm wide.

Flowers: Sweetly honey-scented, brilliant deep-orange. Borne in masses at ends of branches, hiding the foliage. Flowering November – December.

Fruit: Dry, three-winged, 2 – 3 cm wide, dispersed by wind.

Uses: The young roots are peeled and eaten. They have a moist brittle centre that tastes like sugar-candy, and are prized as food. The soft wood is used to make shields.

Comments: The largest 'mistletoe'. The trunk exudes a sweet sticky gum, which does not seem to have been eaten by humans. The Aboriginal name refers to the edible young roots.

GIANT WATERLILY

Scientific Name: *Nymphaea gigantea*

Aboriginal Name: PINANYI (Halls Creek)

References: 24, 46, 51, 65, 85, 102

Habitat: Ephemeral and perennial billabongs on creeks and rivers.

Size and Form: Aquatic perennial herb with stalks up to 2 m long.

Leaves: Floating. Heart-shaped, 30 cm in diameter. Reddish underside, mottled green above.

Flowers: Held above water. Blue with yellow centre, up to 22 cm in diameter. Sweetly scented. As flowers mature, petals and sepals fold back over ovary.

Fruit: Submerged. Urn-shaped, 3 cm wide, with numerous seeds.

Uses: The seed heads and peeled stalks are eaten raw. The roots are roasted in the fire and eaten. The leaves are rubbed on the body to prevent leech bites.

Comments: The 'Uses' for this plant were collected among desert dwellers who had visited Carson River where the plant occurs.

YELLOW PLUM

Scientific Name: *Opilia amentacea*

Aboriginal Name: EMIRRUWA (Arnhem Land)

References: 2, 54, 102, 117, 126

Habitat: Open forest areas and stream margins.

Size and Form: Shrubby parasitic climbing vine or liana. Attaches its roots to those of surrounding vegetation by means of structures known as 'haustoria'.

Leaves: Thin, oval, tapering towards tip, up to 8 cm long.

Flowers: Star-shaped, five-petalled, green, 2 mm in diameter. Borne on small catkin-like flower-heads, which emerge from dormant buds.

Fruit: Ovoid berries, 2 cm long, white flesh. Ripening during early wet season.

Uses: The fruits, which are slightly bitter, are eaten raw.

Comments: This species is widespread throughout the Kimberley and other parts of tropical Australia, however, there is no record of its use in Western Australia.

DESERT WALNUT

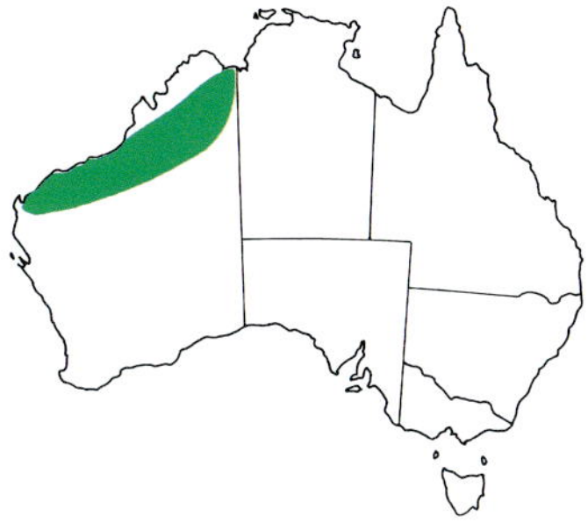

Scientific Name:	*Owenia reticulata*
Aboriginal Name:	BANDAL (Port Hedland)
References:	12, 13, 47, 98
Habitat:	Sandy areas with spinifex cover. Often on dune crests.
Size and Form:	Small tree with dark furrowed bark, up to 8 m tall. Young branches covered in yellow resin, giving a varnished appearance. Dense crown.
Leaves:	Dark-green oval leaflets, 20 cm long, 10 cm wide. Arranged along central rib.
Flowers:	Tiny, 4 – 5 mm in diameter, yellowish, in small sprays near ends of branches.
Fruit:	Globular or pear-shaped, 3 – 4 cm in diameter. Hard, red, with a single large seed.
Uses:	The fruit is roasted, and the kernel removed while hot and rubbed on small skin sores, scratches and insect bites. The fruit, bark and leaves are boiled together and the infusion used as a rub to relieve the symptoms of rheumatic aches and pains.
Comments:	The dense crown of leaves make this a useful shade tree.

MARBLE TREE

Scientific Name: *Owenia vernicosa*

References: 58, 65, 87, 92, 102.

Habitat: Sandy soils in open forest.

Size and Form: Small deciduous tree, up to 10 m tall. Straight thin trunk. Bark loose, thin, almost papery. Branches thick but light and brittle.

Leaves: Compound, composed of 12 – 28 leaflets. Each narrow, sharply pointed, with prominent veins, 12 cm long, 3 cm wide. Clustered at ends of branches.

Flowers: Tiny, white, in spikes at ends of branches.

Fruit: Usually 1 – 2 cm long and pendulous. Initially red, maturing brown.

Uses: The bark is used as a fish poison. Heated leaves are used to relieve sore eyes and headache. A decoction of the bark in water is splashed or poured onto the forehead as a headache cure. The fruit is considered poisonous to humans. It is used for catching animals such as possums, which eat the fruits and become sluggish and easy to catch.

SCREW PALM

Scientific Name:	*Pandanus spiralis*
Aboriginal name:	GUNGURR (Arnhem Land)
References:	6, 12, 62, 65, 87, 102, 105, 126
Habitat:	Open forest, swamps, river banks and foredunes on beaches.
Size and Form:	Palm-like, up to 10 m tall. Trunk narrow, spirally twisted. Bark absent. Older trees often branched, with large prop-like roots supporting the stem base.
Leaves:	Sword-like, up to 2 m long. Small thorns along edges and midrib. Spirally arranged.
Flowers:	White. In clusters, surrounded and obscured by white leaf-like bracts.
Fruit:	Resembling a large woody pineapple, composed of closely packed segments, up to 20 cm in diameter. Deep-orange when mature. Containing numerous small oval kernels clustered tightly around central core.
Uses:	The fruits are roasted and the kernels extracted and eaten. The central part of the upper stem is eaten raw to treat stomach pains. A decoction of the same part is used to treat mouth, tongue and gum soreness.
Comments:	The basal section of each fruit segment is chewed as a sweet food, but it is somewhat irritating to the mouth.

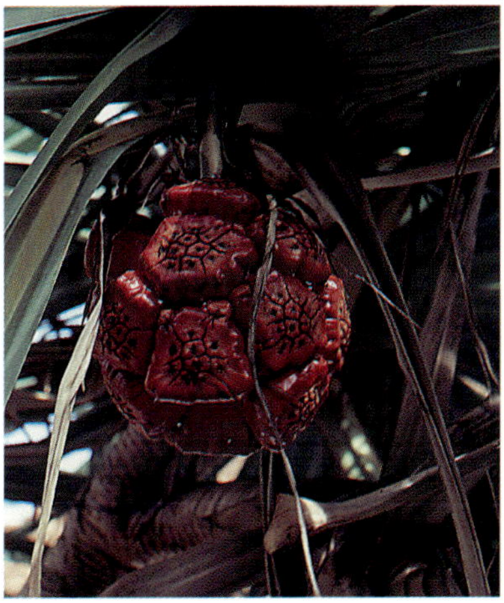

WESTERN WONGA VINE

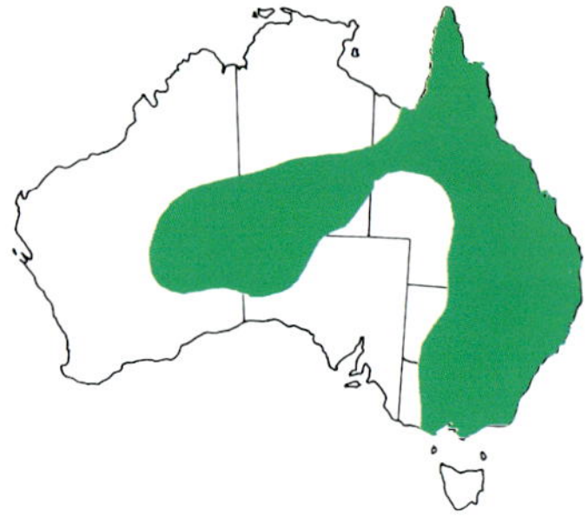

Scientific Name:	*Pandorea pandorana*
Aboriginal Name:	(Not collected)
References:	16, 43, 54, 62, 74, 107
Habitat:	In gorges and on sheltered hillsides. Among rocks (sandstone, quartzite, granite, basalt) in inland areas.
Size and Form:	Shrubby climber with numerous stems, up to 4 m tall, rather untidy.
Leaves:	Composed of 5 – 11 leaflets. Narrow, tapering towards tip, 1.5 – 4 cm long.
Flowers:	Tubular, 1 – 2.5 cm long, borne in large loose clusters. Cream or yellow outside, darker and hairy within.
Fruit:	Beaked pods, 7 – 10 cm long. Seeds flat, oval, 10 mm long, wings 5 – 6 mm long.
Uses:	The stems are cut, tempered in slow fires and used for spear shafts, which are tipped with a section of hardwood, often *Acacia* spp.
Comments:	Previously named *Pandorea doratoxylon*.

KARRI WATTLE

Scientific Name: *Paraserianthes lophantha* subsp. *lophantha*

Aboriginal Name: (Not collected)

References: 71

Habitat: Winter-damp depressions, mainly in jarrah and karri forests.

Size and Form: Shrub or tree, up to 10 m tall. Single trunk when young.

Leaves: Feathery, bipinnate with 8 – 10 pairs of pinnae and 20 – 40 pairs of pinnules.

Flowers: Greenish or yellow 'bottle-brush', up to 8 cm long.

Fruit: Flattish oblong pods, up to 8 cm long, 1 cm wide. Dark-brown, containing six or seven black seeds.

Uses: The leaves contain saponins and can be rubbed to a rich lather with a little water. Useful as 'bush soap'.

Comments: There are other subspecies outside the Perth region. Pods of many *Acacia* spp. are also used in this way.

NONDA PLUM

Scientific Name: *Parinari nonda*

Aboriginal Name: PORTJU (Arnhem Land)

References: 25, 77, 105, 106

Habitat: In dunes behind sandy beaches.

Size and Form: Tree, up to 3 m tall, with hairy branchlets.

Leaves: Oval, up to 8 cm long, 4 cm wide. Densely hairy underside.

Flowers: Tiny, brown-yellow, in clusters at ends of branches.

Fruit: Yellow, egg-shaped, 4 cm long, 2 cm in diameter, with rough kernel.

Uses: The fruit is edible.

Comments: The fruit is astringent if eaten from the tree. It ripens on the ground and is collected direct from the ground for eating.

STINKING PASSION FRUIT

Scientific Name: *Passiflora foetida*

References: 2, 13, 36, 65, 87, 102.

Habitat: Along creek banks, river margins and areas of lush foliage throughout tropical areas.

Size and Form: Scrambling vine with pale-green leaves forming thick tangles.

Leaves: Three-lobed, 7 cm long, hairy, soft.

Flowers: Flattish with several rings of structures, 5 cm wide. White with purple centre, or pale pink with deeper pink centre.

Fruit: Oval, 2 – 3 cm in diameter, soft-skinned. Yellow when ripe. Enclosed in deciduous green 'net', which dries brown.

Uses: The fruit is edible; the pulp is squeezed from the papery case directly into the mouth. An infusion of the leaves is applied for headache or itching skin. To cure ringworm, a decoction of the leaves is applied to skin roughened with *Ficus opposita* leaves.

Comments: An introduced species. The flowers and seeds look similar to those of the cultivated passion fruit, *Passiflora caerulea*. The fruit is quite tasty but strong-smelling. The vegetative parts may be toxic.

QUINE TREE

Scientific Name:	*Petalostigma pubescens*
Aboriginal Name:	JILGARREE (Kalumburu)
References:	12, 24, 60, 87
Habitat:	Open forest. Dry sandy or rocky coastal areas.
Size and Form:	Shrub or small tree, up to 6 m tall. Spreading, umbrella-shaped. Single trunk.
Leaves:	Oval, 5 cm long, 3 cm wide. Dark-green upper surface, densely hairy underside.
Flowers:	Yellow or white. Male flowers composed of 4 – 6 segments, 2 – 4 mm long, clustered in leaf forks. Female flowers solitary, composed of eight segments.
Fruit:	Spherical, up to 2.5 cm in diameter. Yellow, orange or red. Hairy when young.
Uses:	An infusion of the bark or fruit in water is used to relieve sore eyes and toothache, or as a general antiseptic. The bark and fruits are bitter to the taste. The fruit is reputedly useful in malaria treatment, although there is no incidence of malaria in Australia at present.

MALARIA BUSH

Scientific Name: *Petalostigma quadriloculare*

Aboriginal Name: (Not collected)

References: 24, 60

Habitat: Dry, sandy or rocky coastal areas.

Size and Form: Shrub, less than 1.5 m tall. Leaves dark-green, bark deeply furrowed.

Leaves: Oval, 6 cm long, 3.5 cm wide. Margins sometimes toothed. Silky underside. Borne singly at different heights on the axis.

Flowers: Yellow. Male flowers composed of 4 – 6 segments, up to 4 mm long, borne in leaf forks. Female flowers composed of eight segments, up to 4 mm long, solitary.

Fruit: Flattened, globular, 2.5 cm in diameter. Bitter. Orange-yellow when ripe.

Uses: The fruits are held in the mouth to treat toothache. For sore eyes, the bark is put in water and a drop of the infusion added to the eye.

Comments: The fruit was chewed as a remedy for fever and malaria, although there is no incidence of malaria in Australia at present.

TROPICAL REED

Scientific Name: *Phragmites karka*

Aboriginal Name: GAMANGERUNG (Kununurra)

References: 4, 24, 26, 27, 32, 51, 58, 96

Habitat: Common on the fringes of water sources.

Size and Form: Reed, usually 1.8 – 2.1 m tall. Forming compact tussocks up to 0.9 m in diameter, or large dense groves. Underground rhizome.

Leaves: Stems cane-like, up to 1.2 cm wide, smooth, unbranched. Leaf-blade long, tapering gradually to a point. Often with a white powdery ring near the nodes.

Flowers: Arranged in spikes, 13 – 19 mm long, usually with 3 – 6 loose florets. Bracts tapering to a point.

Fruit: Narrow, oblong, pale, smooth. As in other grasses, enclosed within flower parts.

Uses: The stems are used for making bags and baskets. Composite spear shafts are made by inserting a foreshaft of light wood into the reed shaft, which is bound with sinew and fixed with gum cement.

Comments: *Phragmites australis*, common reed, may be confused with this species, but it is not known to be native to Australia and is usually more robust.

MALLEE RICEFLOWER

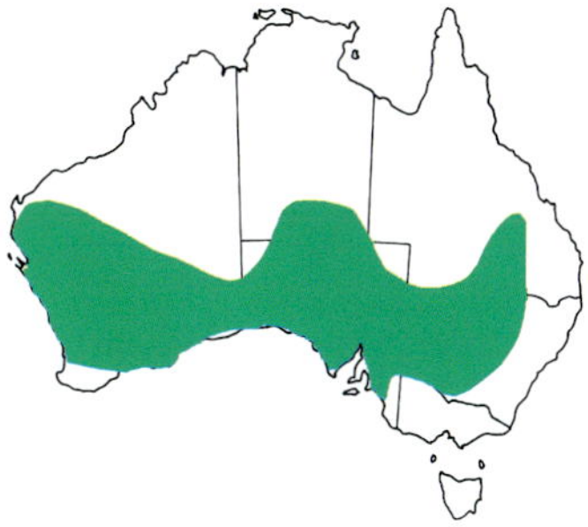

Scientific Name: *Pimelea microcephala*

Aboriginal Name: GUNDAGURRIE (Murchison)

References: 13, 16, 17, 36, 51, 60, 72, 76, 78, 93

Habitat: Usually on sandy or limestone soils in inland areas.

Size and Form: Small much-branched shrub, up to 5 m tall. Hairless except for flowers.

Leaves: Narrow, tapered at both ends, pointed, sometimes blunt, up to 2.5 cm long. Borne singly at different heights on axis.

Flowers: Flower heads 8 mm long, occurring at ends of branchlets, with 2 – 4 leaf-like bracts.

Fruit: Fleshy yellow-green or red berries. Seeds black and wrinkled.

Uses: A decoction of the root bark is drunk for throat and chest complaints. The sweet berries are reputed to be eaten.

Comments: The inner bark of shrubs of this group is very suitable for string making. Fishing line and fine-meshed nets are made with this fibre.

WEEPING PITTOSPORUM

Scientific Name: *Pittosporum phylliraeoides*

Aboriginal Name: KUMPALPA (central Australia)

References: 13, 16, 72, 73, 74, 78, 127, 93, 126

Habitat: Widespread in many habitats.

Size and Form: Attractive weeping shrub or small tree, up to 10 m tall.

Leaves: Oblong, narrow, tapered at both ends, 5 – 10 cm long, often with hooked point. Deep green. Borne singly at different heights on the axis.

Flowers: Tubular, 6 – 12 mm long, yellow. Either solitary or in small short-stalked clusters in leaf forks.

Fruit: Yellow-skinned compressed egg-shaped capsules, 6 – 20 mm long. Splitting open longitudinally into two halves when mature, to reveal several sticky orange to dark-red seeds.

Uses: Infusions of the seeds, fruit pulp, leaves or wood are taken internally to relieve pain and cramps. The fruit is highly astringent.

Comments: Reports from the Nullarbor Plain describe the white seeds being cleaned of their red covering and ground to a flour between two stones. This flour was cooked as a damper.

COCKY APPLE

Scientific Name:	*Planchonia careya*
Aboriginal Name:	WADU (Kununurra)
References:	45, 60, 65, 87, 98
Habitat:	Open forest, ocean dunes and flat areas.
Size and Form:	Shrub or small tree, up to 15 m tall. Bark rough, brown to light-grey, inner bark red and stringy.
Leaves:	Rounded, 15 cm long, 10 cm wide. Waxy. Bright green, or deep-red at certain times of year.
Flowers:	Conspicuous, up to 8 cm wide. Composed of four long green-white petals. Petals folded backwards, revealing large stamen bundle with white tip and pink base. Flowers only live a few hours.
Fruit:	Oval, up to 6 cm long. Green, persistent calyx and pistil. Seeds embedded in cream-coloured fibrous pulp. Ripening soon after flowering.
Uses:	The fruit is eaten raw. The bark is used to make an antiseptic for sores and wounds. Fibres from the roots are used for making string. Roots and bark are mashed and used as a fish poison.
Comments:	The wood is reputed to be used to make fire by friction.

KAHNO

Scientific Name:	*Platysace cirrosa*
Aboriginal Name:	KAHNO (Koorda)
References:	56, 76, 83
Habitat:	Widespread (but not often observed) on laterite and kaolin clay breakaways.
Size and Form:	Slender, often inconspicuous, herbaceous climber, up to 40 cm tall. Main taproot a series of bead-like tubers, 2 – 15 cm in diameter, which become progressively larger with depth.
Leaves:	Linear, stalkless, 2 – 3 cm long. Sparse. Borne singly at different heights on the axis.
Flowers:	Tiny, cream, petals folded inwards. Sepals small. Produced in autumn.
Fruit:	Inconspicuous, circular, laterally compressed. Falls to the ground without opening when ripe.
Uses:	The tubers are useful as food. They are soft, spongy or crisp in texture, according to the time of year that they are harvested, and are rather like dark potatoes. The taste is somewhere between that of coconut and carrot.
Comments:	This species was lost to the botanic world from 1845 until 1974, when the Western Australian Naturalist's Club again collected it.

ROUND YAM

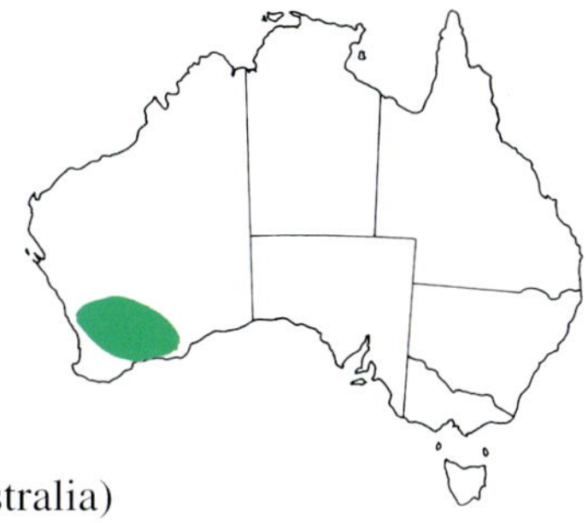

Scientific Name: *Platysace maxwellii*

Aboriginal Name: KARNO, CONNA (south-west Western Australia)

References: 72, 83

Habitat: Common and widespread.

Size and Form: Spreading undershrub, up to 50 cm tall, often tinged with red. Roots tuberous, cylindrical to nearly spherical, 15 – 30 cm long, buried 15 – 20 cm deep.

Leaves: Linear, 10 – 15 mm long, 1 – 2 mm wide. Borne singly at different heights on stem, often restricted to upper parts.

Flowers: Tiny white deciduous corolla. Petals with a prominent midrib. Sepals pale. Ovary smooth, laterally compressed.

Fruit: Disc-like, brown, 2 – 3 mm wide.

Uses: The translucent white flesh of the tubers is eaten raw or roasted. They contain moisture and are sometimes eaten to quench thirst.

Comments: Tubers often occur one below the other, down to a depth of about 50 cm.

SMALL-LEAVED FERN

Scientific Name: *Platyzoma microphyllum*

Aboriginal Name: (Not collected)

References: 65

Habitat: Sandy swampy soil in seasonally wet sites.

Size and Form: Fern, up to 50 cm tall. Forming clumps by means of spreading rhizomes.

Leaves: Infertile fronds fine and grass-like. Fertile fronds divided into tiny leaflets, in pairs on upper part of stem.

Flowers: (None)

Fruit: Spores produced from sporangia on underside of leaves.

Uses: Fertile fronds are used to make children's play necklaces in the West Kimberley.

Comments: At first glance this plant looks more like a rush than a fern, and the thin infertile fronds may be mistaken for grass leaves.

EMU PLUM

Scientific Name: *Podocarpus drouyniana*

Aboriginal Name: KOOLAH (south-west Western Australia)

References: 71, 72

Habitat: Lateritic loams and sandy soils in jarrah and karri forest.

Size and Form: Low shrub, up to 1.2 m tall. Multiple stems arising from mallee-like rootstock.

Leaves: Narrow, grey-green, 5 – 10 cm long.

Flowers: Male and female cones on separate plants.

Fruit: After a bushfire the female cone stalk becomes fleshy and rounded, up to 3 cm long, 2.5 cm wide. A single seed develops inside the fleshy scales.

Uses: The 'fruit' (actually the fleshy stalk or receptacle) is edible, with a mildly resinous flavour. Sometimes used by early settlers for jams and preserves.

Comments: Related to conifers. At maturity the fruit is plum-coloured with a light waxy coating. *Podocarpus elatus* from the brush forests of eastern Australia is a tall tree having similar fruits.

PIGWEED

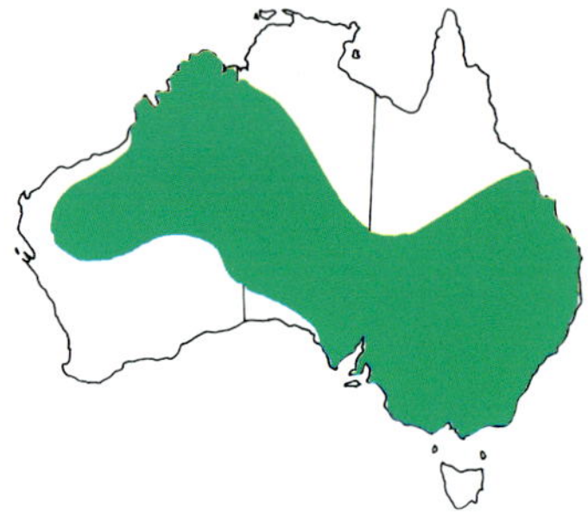

Scientific Name: *Portulaca oleracea*

Aboriginal Name: MUNYEROO (central Australia)

References: 16, 24, 36, 60, 62.

Habitat: Widespread and common in a wide variety of soils. Most abundant where soil has been disturbed.

Size and Form: Succulent annual, with red-brown stems lying flat on the ground.

Leaves: Wedge-shaped or oblong wedge-shaped, sometimes becoming rounded near tip, 1 – 2 cm long. Mostly borne singly at different heights on stems.

Flowers: Yellow, 1 cm in diameter. Stalkless, solitary or in clusters in leaf forks. The 4 – 6 petals are as long as the sepals.

Fruit: Seeds numerous, black, like poppy seeds.

Uses: The leaves are eaten whole as a cooling diuretic. This prevents or cures scurvy, a disease caused by vitamin C deficiency. It has been used by Aborigines and early settlers as a blood cleanser.

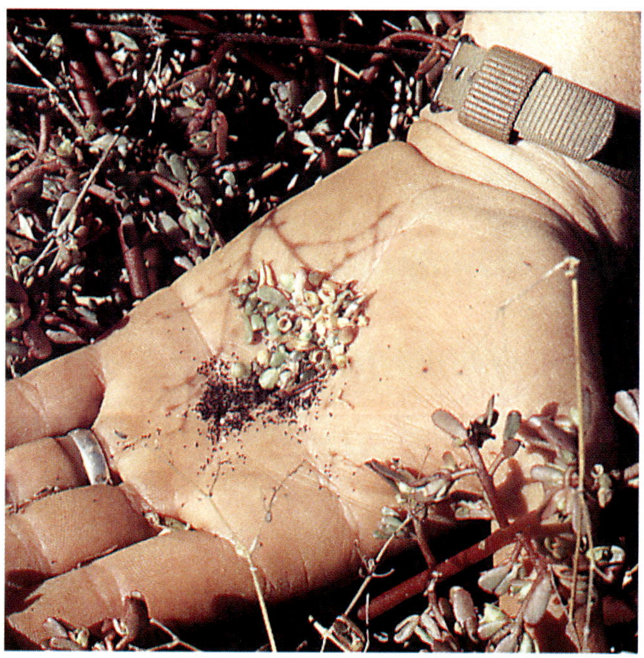

YELLOW BOXWOOD

Scientific Name: *Pouteria pohlmaniana*

Aboriginal Name: NANGI (Kalumburu)

References: 2, 60, 65, 100, 116

Habitat: Open forest near coasts.

Size and Form: Small tree, 4 – 5 m tall. Bark grey and scaly, exuding a milky juice when cut.

Leaves: Dark green, oval, 8 – 15 cm long.

Flowers: Tiny, white, bell-shaped. In clusters in leaf forks.

Fruit: Globular, 2.5 cm in diameter, greenish. Containing five hard brown glossy slightly flattened seeds, 12 mm long.

Uses: Fruit edible when soft (but remains green).

Comments: Most of the flowers drop without producing fruit. Another species, *Pouteria sericea*, produces edible black fruits, which are highly prized as bush tucker.

A LEEK ORCHID

Scientific Name: *Prasophyllum* sp.

Aboriginal Name: TUBOC (south-west Western Australia)

References: 72, 83

Habitat: Widespread, often in seasonally wet sites such as swamps and the margins of granite rocks.

Size and Form: Erect herbaceous perennial, usually less than 1 m tall. Tuber white, ovoid, up to 3.5 cm long, buried 3 – 15 cm deep.

Leaves: Smooth, linear, erect, cylindrical and hollow. Stem enclosed within the leaf for about half the length of the leaf.

Flowers: Obvious 'tongue' typical of orchids. Almost stalkless and crowded in a fairly dense spike. Lateral petals narrower and smaller than the others.

Fruit: Brown elongated capsules containing numerous powder-like seeds.

Uses: Aboriginal women use a digging-stick to collect tubers from this and other orchid groups. They are usually roasted before being eaten.

Comments: Certain horizontal root-like outgrowths swell terminally to produce up to three daughter tubers per plant. The parent tuber is replaced annually.

JOCKEY'S CAP

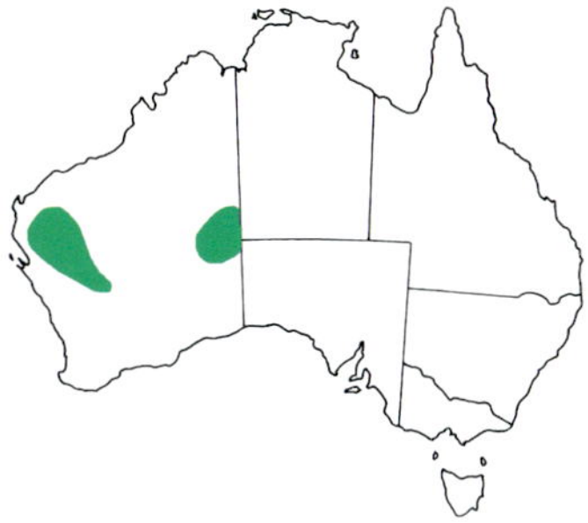

Scientific Name: *Prostanthera striatiflora*

Aboriginal Name: WARLKALPA (Balgo)

References: 51, 74, 76, 97, 126, 127

Habitat: Usually in sheltered places on rocky ranges, often around granite outcrops.

Size and Form: Smooth shrub, 1 – 2 m tall.

Leaves: Narrow, oval, 10 – 25 mm long, 3 – 10 mm wide, apex rounded. Stalkless, leathery, margins slightly curved.

Flowers: White, yellow-spotted in throat, purple striations in tube, 20 – 25 mm long. Sparsely hairy outside.

Fruit: Like small apricots arising from dry flower parts, 15 mm long. Sepals around fruit membranous.

Uses: The leaves are very strongly scented. They are boiled in water, and the liquid used to bathe patients for general illness. Dried leaves are spread on the water in water holes to stun birds.

Comments: The poison is too powerful for the water to be safely used by humans, so branches are always placed beside the waterhole to show that it has been poisoned.

WILD ASPARAGUS

Scientific Name:	*Protasparagus plumosus*
Aboriginal Name:	WAITJAN (Darwin area)
References:	65, 120
Habitat:	Open forest, sometimes semi-swampy areas.
Size and Form:	Thorny scrambling plant with weak stem, up to 3 m tall. Resembling common asparagus of horticulture. Abbreviated rhizome with numerous swollen tubers.
Leaves:	Needle-like, in tufts, often small. Shed in dry season. Sharp thorns where leaf joins stem.
Flowers:	Small, white, clustered on short stalks from main stem.
Fruit:	Small two-lobed berries, 6 mm in diameter. Red when ripe.
Uses:	The swollen fibrous roots are crushed, boiled and applied to the skin for sores and general sickness. Not eaten; believed to be poisonous.
Comments:	Often obscured in forest undergrowth; a thorny trap for the unwary walker.

WIRA

Scientific Name: *Psoralea leucantha*

Aboriginal Name: WIRA (Balgo area)

References: 51

Habitat: Along flood plains and hill slopes.

Size and Form: Erect shrub, up to 2 m tall. Hairless or covered with short soft erect hairs.

Leaves: Composed of 1 – 3 ovate leaflets, tapering towards sharp point, 5 – 7 cm long, 1 – 2 cm wide. Margins smooth, marked with dots.

Flowers: Flower head 40 – 50 mm long. Petals white, dark-tipped, 5 – 6 mm long. Sepals hairless, marked with dots, 3 mm long.

Fruit: Black wrinkled hairless pods, elongated to egg-shaped, 7 mm long, 4 mm wide.

Uses: Children use the stems to make little spears. 'Shaved' sticks, produced by scraping shavings off the stem into rings at about 10 cm intervals, are used to decorate ceremonial artefacts, and as decorative 'hairpins'. These also make good tinder for fire-lighting.

COMMON BRACKEN

Scientific Name: *Pteridium esculentum*

Aboriginal Name: (Not collected)

References: 18, 53, 57, 67, 86, 93, 124

Habitat: Moist open forest and along watercourses. Invasive pasture weed.

Size and Form: Tough brown-stemmed fern, 1 – 2 m tall, occasionally reaching 4 m in shady places. Long creeping rhizome.

Leaves: Broadly triangular. Finely divided into numerous leaflets. Rigid, dark green, glossy or dull. Leathery and smooth on the upper surface, fine flat hairs on the lower surface.

Flowers: Non-flowering. Small lobe where each narrow leaflet joins stalk.

Fruit: Like all ferns, the 'fruits' are spores produced in small structures beneath the fronds. Spore-bearing bodies fairly inconspicuous.

Uses: The starchy rhizomes have been used as food by Maoris in New Zealand.

Comments: The hairy 'fiddleheads' or young shoots may be linked to cancer of the stomach, and should not be eaten. Adult leaves and stems are rich in tannic acid and have been used for tanning hides.

TJUNGARAI

Scientific Name: *Pterocaulon serrulatum*

Aboriginal Name: TJUNGARAI (Katherine)

References: 47, 51, 74, 91

Habitat: Tropical areas.

Size and Form: Tall strongly scented hairy perennial herb, up to 90 cm tall.

Leaves: Broad oval, tapering towards tip, 2 – 5 cm long. Toothed margins, wrinkled blistered surface.

Flowers: Flower heads in oval pinkish clusters up to 1 cm wide, at ends of stems.

Fruit: Small seeds with hairy 'tails', clustered in the dried flower parts.

Uses: The leaves are inserted into the nose for head colds, and chewed to relieve congestion of the lungs. Wounds are treated by rubbing the affected area with leaves.

APPLE BUSH

Scientific Name: *Pterocaulon sphacelatum*

Aboriginal Name: BANDINGUN MANGUNMA (Balgo)

References: 51

Habitat: Sandy gravels along watercourses and rocky hills.

Size and Form: Woolly aromatic herb, up to 65 cm tall, branching stems, somewhat spreading.

Leaves: Oval, up to 35 mm long, grey-green, raised veins. Dense woolly coat of soft white hairs, velvety to the touch. Borne singly at different heights on stem.

Flowers: Spiky green-white flower-heads at ends of branches. Flowering August.

Fruit: Seeds in old flower parts, 10 mm long.

Uses: The whole dry plant is ground to a powder and inhaled as a decongestant. The fresh flowers or twigs are placed in a hole pierced in the nasal septum to relieve congestion. The hairy parts of the plant could be used as tinder for fire-lighting.

Comments: Many aromatic plants are used as decongestants, and to relieve the symptoms of colds and influenza. It is not known when these two illnesses arrived in Australia, but detailed knowledge of plant properties allowed herbal treatments to be developed in parallel with European medicines. For example, compare the use of this plant with that of 'Vick's Vaporub'.

TALL MULLA MULLA

Scientific Name:	*Ptilotus exaltatus*
Aboriginal Name:	TJIRUNNA (Roebourne)
References:	51
Habitat:	Widespread and common, especially in medium-textured and limestone-based soils.
Size and Form:	Short-lived erect herb, growing to 1.5 m tall in good seasons.
Leaves:	Oblong, tapering towards tip, 20 cm long, 7 cm wide. Glossy, dark green, erect.
Flowers:	Spikes purple fading to mauve, 5 – 10 cm long, 3 – 4 cm in diameter, papery texture.
Fruit:	Insignificant. Within old flower parts.
Uses:	This plant marks a section group of Aboriginal people living near Roebourne, and has human 'relatives' amongst the group.
Comments:	It is apparently not grazed by cattle, but is eaten by horses as a last resort.

SILVER BUSH

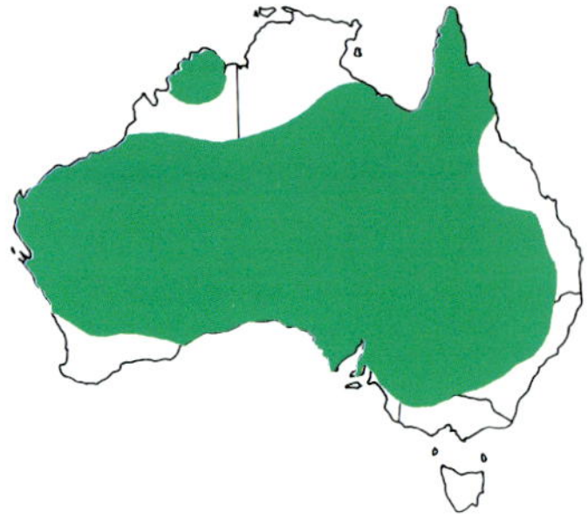

Scientific Name: *Ptilotus obovatus*

Aboriginal Name: MATAKURU (Great Sandy Desert)

References: None found.

Habitat: Widely distributed in arid regions on open stony ground, in shrubland and woodland.

Size and Form: Shrubby perennial, 10 – 120 cm tall and of similar width. Stems and foliage softly hairy.

Leaves: Oval, 4 cm long, 1.5 cm wide.

Flowers: Short cylindrical or hemispherical pale-pink spikes, 1 – 3 cm long. Many on ends of stalks. Flowering most of the year.

Fruit: Insignificant, hidden among dry flower parts.

Uses: The fluffy white flower-heads are beaten into a frothy mass and used for decoration in dancing. They are stuck to the bodies of young tribesmen using blood and gum (or now, sugar solution).

Comments: The dry flower parts make a passable tinder for fire-lighting.

RED MANGROVE

Scientific Name: *Rhizophora stylosa*

Aboriginal Name: BINDUN (Broome)

References: 64, 92, 94, 121

Habitat: Northern coastal areas.

Size and Form: Columnar shrub or tree, up to 20 m tall. Bark grey to brown-black, smooth, fissured.

Leaves: Broadly oval, up to 15 cm long. Glossy green, thick, brittle. Numerous small red-brown glands on undersurface. Clustered towards ends of branchlets, borne at same level on opposite sides of axis.

Flowers: White, up to 15 mm wide, borne in pairs in leaf axils.

Fruit: Brown, shaped like the flame of a match, 3 cm long, 2 cm wide. Long cylindrical 'tail' helps fruit to float, and later acts as an anchor. 'Tail' reaches 35 cm in length and 2 cm in diameter before seeds are released.

Uses: The inner bark is scraped into water to make an infusion, which is used to treat ulcers and yaws. The suckers are boiled in water and the liquid applied to chicken pox blisters.

Comments: Kimberley Aboriginal children make toy rafts by pinning together five or six of the fruit 'tail' structures to make a platform. Two of these are linked together to mimic the larger rafts used by their parents. (See *Camptostemon schultzii*.)

WATERCRESS

Scientific Name: *Rorippa nasturtium-aquaticum*

Aboriginal Name: (Not collected)

References: 75, 124

Habitat: In fresh, often flowing, water.

Size and Form: Herbaceous perennial with many rootlets and trailing stalks, 70 – 80 cm long.

Leaves: Deeply lobed, up to 20 cm long. Green. Terminal leaf largest and rounded. Basal ones having stalks, upper ones stalkless and segmented.

Flowers: Whitish, four petals and four sepals. Sepals (up to 7 mm long) are twice as long as petals.

Fruit: Broad, dry, composed of two parts or 'wings'. Containing seeds arranged in rows on each side.

Uses: The leaves and stems are eaten raw or cooked as a vegetable.

Comments: An introduced species, now naturalised in several localities. Previously named *Nasturtium officinale*.

QUANDONG

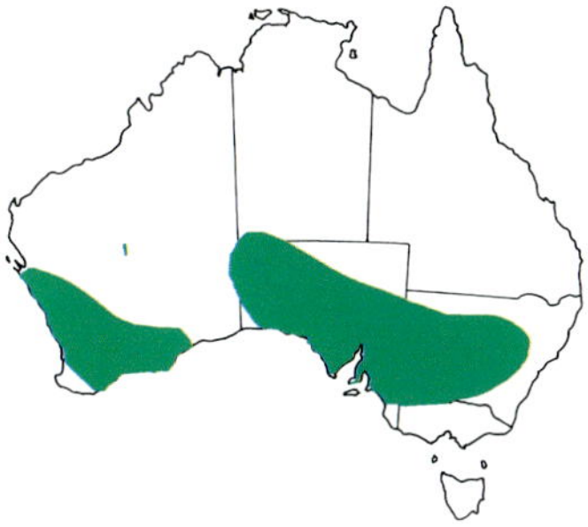

Scientific Name: *Santalum acuminatum*

Aboriginal Name: WOLGOL (south-west Western Australia)

References: 16, 42, 62, 74, 76, 102.

Habitat: Widespread and common, favouring sandhills and sandy plains.

Size and Form: Shapely tree or shrub with a round crown when mature, up to 6 m tall. Bark rough, dark-grey, fissured. Foliage fresh green, drooping.

Leaves: Tapering towards tip and either straight or gently curved, 3 – 9 cm long, 0.3 – 1.5 cm wide.

Flowers: Tiny, pale-green to orange, star shaped. In inconspicuous groups towards the ends of branchlets. Fetid odour.

Fruit: Spherical, up to 2.5 cm in diameter. Thinly fleshed, bright red when ripe. Single large crinkled hard seed.

Uses: The flesh of the fruit is edible when ripe, but rather acid. The hard seed can be roasted and the oily kernel eaten. Pitted fruits can be dried for storage and reconstituted into a paste by grinding with water.

Comments: The yellow flesh and pitted stone of the fruit give this plant the less well-known common name of native peach.

BLACK PLUM

Scientific Name:	*Santalum lanceolatum*
Aboriginal Name:	BILALUR (Broome)
References:	87, 113
Habitat:	Usually on creek banks and gullies. Found on sandplains and spinifex or shrub plains in northern areas.
Size and Form:	Shrub, up to 7 m tall. Bark grey, fissured. Branches and leaves hanging down.
Leaves:	Tapering towards tip, 2 – 9 cm long, 0.5 – 2.5 cm wide. Pale grey.
Flowers:	Tiny, white or yellow, in loose clusters at ends of branches.
Fruit:	Purple, plum-like, 7 – 10 mm in diameter, with prominent circular scar at lower end. Smooth seed.
Uses:	The roots are mashed, soaked in water, and the liquid used as a liniment. The oily seed is ground to a paste with water and used similarly. The fruit is edible. The leaves are burnt to repel mosquitoes. In Queensland, an infusion of leaves and bark is drunk as a purgative.
Comments:	Sometimes referred to as 'sandalwood', but this name is better applied to *Santalum spicatum*.

BITTER QUANDONG

Scientific Name: *Santalum murrayanum*

Aboriginal Name: (Not collected)

References: 51

Habitat: In sand on spinifex or shrub plains.

Size and Form: Shrub, up to 2 m tall. Bark smooth, branchlets pendulous.

Leaves: Narrow, tapering towards tip, curved, 15 – 35 mm long, 1.5 – 4 mm wide. Pale green. Often in groups of three.

Flowers: Four yellowish petals, the whole 3 mm wide, on a tube 1 mm long. Many flowers in large clusters.

Fruit: Spherical, 15 – 23 mm in diameter. Apical scar left by perianth. Green or brownish, shell of kernel deeply wrinkled.

Uses: The fruit has bitter and inedible flesh. The kernels can be roasted in the fire ashes and eaten.

Comments: The oily kernels may have been rubbed onto wooden artefacts to assist in their preservation.

SANDALWOOD

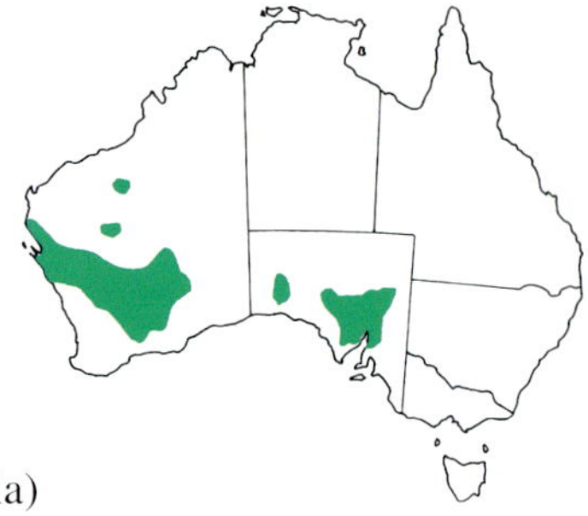

Scientific Name:	*Santalum spicatum*
Aboriginal Name:	POILYENUM (south-west Western Australia)
References:	47, 51, 72, 76, 78
Habitat:	Areas of red loam with mulga, or skeletal soils on granite outcrops.
Size and Form:	Shrub, up to 4 m tall. Bark rough, grey. Branchlets stiff, spreading.
Leaves:	Broad, tapering towards tip, or narrow oval with rounded tip, 2 – 7 cm long, 0.3 – 1.5 cm wide, grey-green.
Flowers:	Scented, red-green, 3 mm wide. Squarish, with four petals. Borne in clusters.
Fruit:	Spherical, 1.5 – 2 cm in diameter. Outer skin green or brown, inner skin smooth.
Uses:	The nuts have a white kernel, which produces a good quality oil. Although edible, the nuts are not eaten by Aborigines, but are crushed and used as a liniment on aching joints.
Comments:	Sandalwood trees are 'pulled', roots and all, from the ground and exported for the production of incense. A replanting program has been initiated in Western Australia. The smoke is thought to induce nausea.

CAUSTIC VINE

Scientific Name:	*Sarcostema australe*
Aboriginal Name:	IIPI-IIPI (Balgo)
References:	13, 16, 47, 60, 102
Habitat:	Widespread but sporadic in dry areas, on stony soils inland and near coasts. Also occurs on margins of vine thickets and in savanna in the Kimberley region of Western Australia.
Size and Form:	Straggling shrub, up to 1.5 m tall, or sometimes twining to the top of trees. Stems smooth, jointed.
Leaves:	Foliage minute.
Flowers:	Tiny, cream-white, waxy, star-shaped, in clusters in leaf forks.
Fruit:	Pod-like, pointed, 5 – 8 cm long. Splitting lengthwise to reveal many silky flat seeds.
Uses:	The sap is applied directly to scabies and itchy sores. It is best to gather the sap after rain, when the sap flow is stronger. Infusions of the whole plant are used similarly.
Comments:	When wounded, the twigs exude copious quantities of white corrosive sap.

PRICKLY FANFLOWER

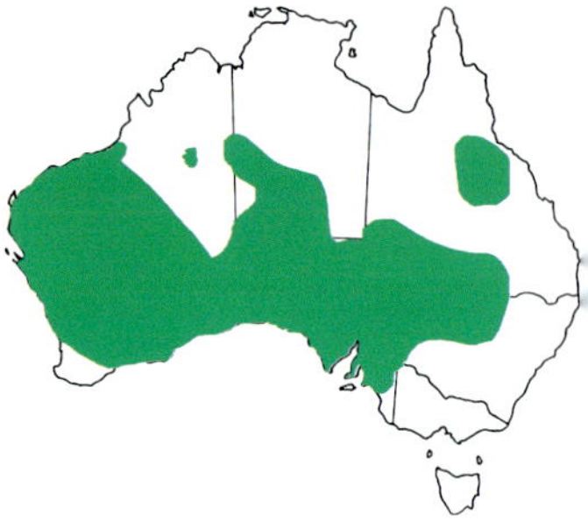

Scientific Name:	*Scaevola spinescens*
Aboriginal Name:	GUBARU (Mt Magnet)
References:	6, 18, 60, 62, 81, 82
Habitat:	Sandy loams in inland regions.
Size and Form:	Rigid scrubby bush, 1 – 2 m tall, covered with fine hairs.
Leaves:	Narrow, broadened near the rounded tip, 10 – 25 mm long. Borne singly or in clusters at different heights on the axis.
Flowers:	Each bush carries a few solitary white-yellow flowers (sometimes bronze coloured) on short slender stalks in leaf axils.
Fruit:	Black or purplish drupes, up to 8 mm in diameter.
Uses:	Decoctions made from the roots are drunk for stomach ache and urinary troubles. A decoction of the broken stems is also used as a wash to cure sores and rashes. The ripe fruit is edible raw.
Comments:	This plant has been at the centre of a controversy over its possible use as a cancer cure. Known as 'maroon bush', it is reputed to assist in cases of stomach cancer, but has not been tested under controlled conditions. Casual experimentation is not recommended.

LIMESTONE SENNA

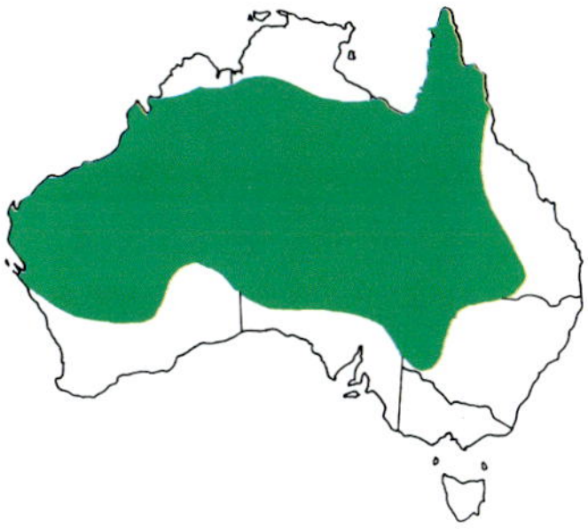

Scientific Name: *Senna artemisoides*

Aboriginal Name: PUNTILIRRI (Balgo)

References: 1, 28, 51, 113

Habitat: Grasslands and plains adjacent to watercourses and rocky outcrops.

Size and Form: Shrub, 1 – 2 m tall. Rounded or flat-topped, often spreading. Glaucous, covered with short soft hairs.

Leaves: Compound, 5 – 6 cm long, composed of 1 – 3 pairs of leaflets borne on opposite sides of stem. Leaflets oval, tapering at base, up to 2 cm long, lower pairs smaller. Generally firm and leathery. Stalks 0.5 – 2 cm long.

Flowers: Yellow, buttercup-like, in clusters of 6 – 20 flowers.

Fruit: Flat, oblong or curved pods, 7 cm long, 1.5 – 2.5 cm wide. Containing 12 flattish oblong dark dull seeds.

Uses: The green leaves are boiled in water and the resulting red infusion used as a 'wash' for fevers.

Comments: Formerly known as *Cassia oligophylla*. Senna bark is known to be a strong laxative, and the leaves of some species have also been used in this manner. Ingestion of the plant is not recommended.

COCKROACH BUSH

Scientific Name: *Senna notabilis*

Aboriginal Name: YARINUNTA (Roebourne)

References: 1, 109, 113

Habitat: Sandplains and gravelly soils. Favours disturbed areas, and is used in re-vegetation projects.

Size and Form: Small dense short-lived shrub, up to 1 m tall.

Leaves: Up to 15 cm long, composed of up to 13 pairs of leaflets. Leaflets tapering towards tip, 2.5 cm long, 1 cm wide. Covered with long coarse hairs giving the plant a greyish appearance.

Flowers: Clusters of yellow flowers each about 2 cm diameter, each with five petals. occur at tips of branches.

Fruit: Oval pods, flat but ridged where the dark seeds occur, 3.5 cm long, 1.5 cm wide. Turning orange when seeds ripen.

Uses: The seeds are eaten by birds.

Comments: Formerly known as *Cassia notabilis*. Senna bark is known to be a strong laxative, and the leaves of some species have also been used in this manner. Ingestion of the plant is not recommended.

CANDLESTICK SENNA

Scientific Name:	*Senna venusta*
Aboriginal Name:	(Not collected)
References:	1, 87, 109, 113
Habitat:	Scrub country, watercourses and rocky sandstone hills in low woodland.
Size and Form:	Shrub, up to 1 m tall.
Leaves:	Showy, up to 25 cm long. Composed of 15 pairs of oblong leaflets, each 4 cm long, 2 cm wide.
Flowers:	Yellow, with five petals. Borne in large terminal flower-heads, which give the plant its common name.
Fruit:	Cylindrical pods, 10 cm long. Containing oblong, dark-brown seeds, up to 7 mm long.
Uses:	The leaves and twigs are boiled and the resulting sticky liquid is applied to sores, itchy areas and ringworm. It should not be applied to the eyes or mouth.
Comments:	Formerly known as *Cassia venusta*. Senna bark is a strong laxative, and the leaves of some species have also been used in this manner. Ingestion of the plant is not recommended.

WHITE DRAGON TREE

Scientific Name: *Sesbania formosa*

Aboriginal Name: PIDUNGARRA (Roebourne)

References: 47, 51, 76, 102

Habitat: Alluvial plains and river banks.

Size and Form: Fast-growing medium-sized tree, 8 – 12 m tall. Bark rough, corky.

Leaves: Compound, up to 40 cm long, with 20 pairs of opposed leaflets, each 3 – 6 cm long.

Flowers: Large, cream-white, peaflower-like, 8 cm long.

Fruit: Thin pods, 50 cm long, containing numerous brown seeds.

Uses: The ash is used as a topical dressing on wounds. A decoction of the leaves is used to wash or colour hair. The flowers are edible and are used in Asian cooking.

KANGAROO APPLE

Scientific Name:	*Solanum aviculare*
Aboriginal Name:	WEAKICH (Lake Condah)
References:	63, 86
Habitat:	Widespread in wet eucalypt forest and near rainforest margins.
Size and Form:	Perennial shrub, up to 2 m tall.
Leaves:	Dark green, 10 – 30 cm long. Most lower leaves deeply divided into long tapered triangular segments. Stalks 3 cm long.
Flowers:	Blue, violet or purple, up to 3.5 cm in diameter. Five-sided, with sharply pointed lobes. In open bunches in leaf forks.
Fruit:	Smooth egg-shaped berries, orange-red to yellow, 2.5 cm long, on stalks. Containing numerous seeds in a soft pulp.
Uses:	When fully ripe the fruit is eaten raw, or baked until the skin ruptures.
Comments:	This species is a source of steroid drugs. It has been introduced into gardens in southern Australia as a horticultural subject, but is of dubious decorative value.

DESERT RAISIN

Scientific Name: *Solanum centrale*

Aboriginal Name: KAMPURARA (Balgo)

References: 16, 22, 76, 84, 123, 127

Habitat: Red sandhill country, often in areas burnt in the previous season.

Size and Form: Low shrub, up to 45 cm tall. Soft furry stems and leaves, some plants with prickles on the stems.

Leaves: Oval, 4 cm long, 1.5 cm wide. Light-green when young, changing to olive-green.

Flowers: Corolla a broad five-pointed 'star'. Pale or deep purple. Often on stalks which are up to 1 cm long and bent sharply downwards. Arranged in sprays composed of 1 – 6 flowers.

Fruit: Globular, up to 1.5 cm in diameter. Produced in clusters under leaves. Ripening from purple-green to green-white to yellow-brown. Wrinkled and raisin-like.

Uses: The fruit is eaten raw when green-white to yellow-brown. When dried and shrivelled up, it resembles and tastes like sultanas.

Comments: The dried fruits last quite well. Sometimes they were pounded together between grinding stones to make a ball of fruit pulp, which could be reconstituted with a little water to make a jam-like substance.

RIVER BUSH TOMATO

Scientific Name: *Solanum diversiflorum*

Aboriginal Name: KARLUMPU (Roebourne)

References: 12, 13

Habitat: Widespread, but most common in drainage lines and gravel soils in sheltered locations.

Size and Form: Small shrub, 15 – 50 cm tall. Prickly stems and usually leaves.

Leaves: Ovate to oblong, 30 mm long, 10 – 20 mm wide. Divided into lobes 2 – 5 mm long. Lobes towards tip not cut to midrib.

Flowers: Spray of up to 20 small purple male flowers above a bisexual pentagonal flower, 3 cm in diameter, purple with bright-yellow centre

Fruit: Spherical, firm, 1.5 – 3 cm in diameter. Outer portion pale green-yellow, turning white when ripe. Seeds dark brown-black.

Uses: The fruit is cooked in ashes and the seeds scooped out before the fleshy outer parts are eaten.

Comments: A large-fruited *Solanum*. Small paddle-shaped wooden spoons are made especially for cutting open the fruit and scooping out the pulp.

POTATO BUSH

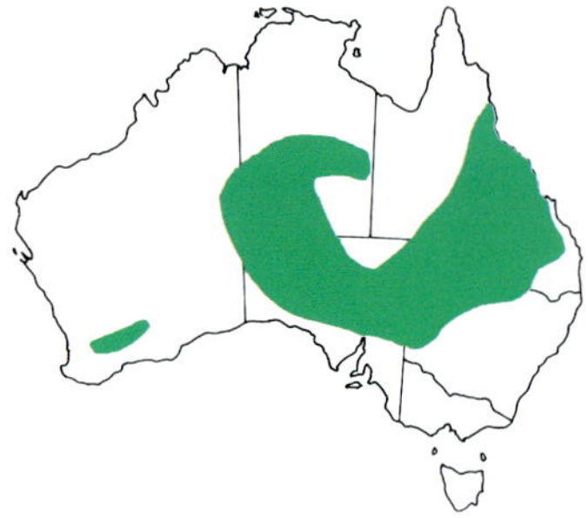

Scientific Name:	*Solanum ellipticum*
Aboriginal Name:	YURALPA (Balgo)
References:	61, 62, 74, 78, 83, 123.
Habitat:	Sandhill country, at the foot of rocky hills and on slopes.
Size and Form:	Sprawling shrub, up to 40 cm tall. Prickles on stems and crown of fruit.
Leaves:	Narrow, oval, 6 cm long, 2 – 3 cm wide. Margins smooth or wavy. Stalks 1 – 5 cm long. Grey-green.
Flowers:	Short tube and spreading star-shaped lobes, 2 – 3 cm in diameter, purple. Anthers 4 mm long. Arranged in sprays composed of 1 – 6 flowers. Stalks 1 – 8 cm long.
Fruit:	Globular or slightly ovoid berries, 1.5 – 2 cm in diameter. Pale yellow-green, often flushed with purple. Seeds pale, 2 – 2.5 mm long.
Uses:	The fruits are eaten raw and have a bitter-sweet juicy taste.
Comments:	This plant has mythological associations among desert Aborigines, who conduct special ceremonies to ensure its proliferation.

BLACKBERRY NIGHTSHADE

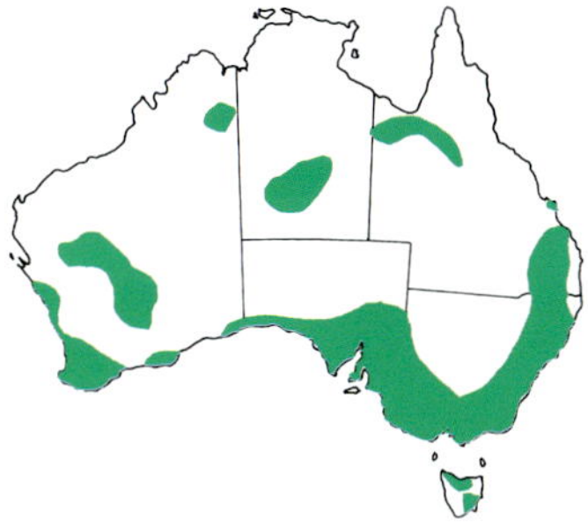

Scientific Name: *Solanum nigrum*

Aboriginal Name: (Not collected)

References: 57, 67, 72, 75, 78, 102, 127

Habitat: A widespread introduced weed of cultivated lands.

Size and Form: Annual, shrub-like, up to 1 m tall. White taproot.

Leaves: Slender, oval to diamond-shaped, pointed, slightly lobed margins, up to 8 cm long. Mid-green, sparsely covered with short soft hairs. Borne singly at different heights on the axis.

Flowers: White, five-lobed, up to 10 mm in diameter. Borne in sprays on short stems.

Fruit: Peppercorn-sized berries. Green at first, becoming juicy and black as they ripen. Containing many small kidney-shaped seeds.

Uses: Ripe fruit is reputedly edible. Eating green fruits causes nausea, vomiting and diarrhoea.

Comments: Recent investigations reveal that there are several varieties of Blackberry Nightshade in Australia and the fruits of some of these never lose their toxicity.

A BUSH TOMATO

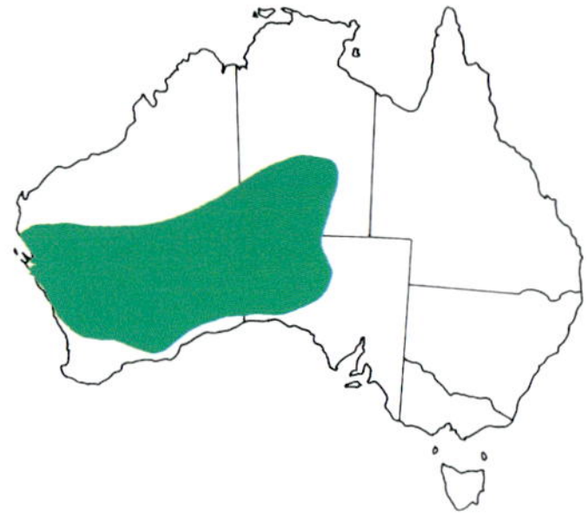

Scientific Name: *Solanum orbiculatum*

Aboriginal Name: (Not collected)

References: 51, 82

Habitat: Usually on sand.

Size and Form: Erect or rounded shrub, 0.5 – 1.5 m tall. Prickles on stems up to 1 cm long.

Leaves: Broadly oval, rounded tip, 15 – 60 mm long, 15 – 35 mm wide. Smooth margins, stalks 5 – 10 mm long.

Flowers: Violet, lobes tapering towards tip, 20 – 25 mm in diameter. Anthers 6 – 7 mm long, ovary smooth, style erect. Arranged in sprays composed of 1 – 4 flowers, on stalks less than 1 cm long.

Fruit: Fruiting flower stalks usually bent downwards. Berries in a prickly calyx, 1.5 cm in diameter. Ripening to pale yellow-cream. Seeds succulent, 3 mm long, pale or light brown.

Uses: The yellow fruit is eaten.

Comments: There are two subspecies of this plant. The fruits of both are probably edible.

AUSTRALIAN BLUEBELL

Scientific Name: *Sollya heterophylla*

Aboriginal Name: GUMUG (Katanning)

References: 12, 54

Habitat: Hardy, common in woodland areas.

Size and Form: Rambling shrub, 3 – 4 m tall, sometimes scrambling through adjacent trees and shrubs.

Leaves: Elongated oval, tapering towards tip, 6 cm long, bright green.

Flowers: Five broad petals, tapering towards tip. Borne in drooping clusters at ends of branchlets. Usually bright blue.

Fruit: Fleshy, blue-black, turning translucent whitish when ripe. Hanging downwards. Produced in abundance.

Uses: The fruits are edible.

Comments: A popular garden plant in Western Australia, growing in two forms: either a small rounded shrub, or a rambling climber.

STICKY STEMODIA

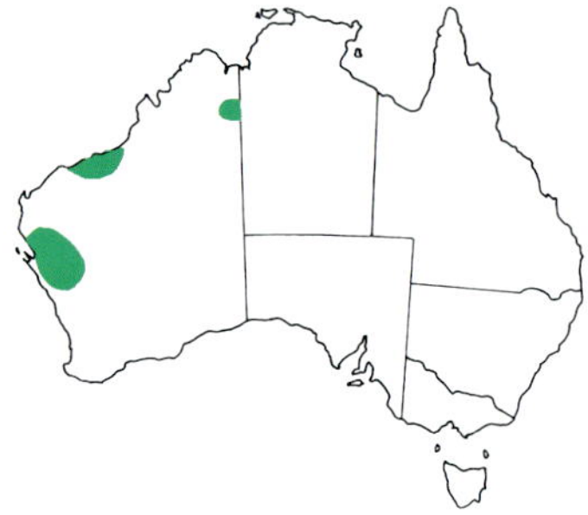

Scientific Name: *Stemodia viscosa*

Aboriginal Name: PONJA-PONJA (Balgo)

References: 13, 93, 97, 127

Habitat: Moist ground and river banks.

Size and Form: Perennial herb, up to 50 cm tall. Sticky, hairy, aromatic.

Leaves: Roughly oval, 3 – 6 cm long. Irregularly toothed margins. Borne at same level on opposite sides of stem.

Flowers: Tubular, with separate upper and lower lips, 2 cm long. Blue. Lower lip yellow at base. Occurring in leaf forks. Flowering autumn and winter.

Fruit: Ovoid capsules, 5 mm long, 2 mm wide. Enclosed in remnants of flower parts and tapering to a point. Containing yellow-brown seeds.

Uses: The bruised plant is used as a remedy for colds and rubbed or sniffed to prevent sickness. Also used to make beds for babies to keep them healthy.

Comments: This plant also occurs in India, and is used there to relieve the symptoms of colds and respiratory complaints.

PEANUT TREE

Scientific Name: *Sterculia quadrifida*

Aboriginal Name: DALU (Kalumburu)

References: 113

Habitat: A variety of habitats: vine thickets, sand dunes behind beaches and rocky slopes.

Size and Form: Tree, up to 12 m tall. Straggly, smaller stems thick. Bark smooth, grey.

Leaves: Elongated heart shape, 23 cm long, 17.5 cm wide. Dark green, rough, prominent veins. Deciduous, leaves shed in June or July, new growth in October – November.

Flowers: Small and inconspicuous. Borne in terminal flower-heads. Sepals remaining joined at tips, forming a tubular 'cage' up to 1 cm long. Outer surface of sepals green or yellow and hairy.

Fruit: Four or five pods carried stiffly on stems in a radiating form. Skin tough, leathery, red when ripe. Ripe pods splitting to reveal white inner surface and black seeds. Seeds up to 1 cm long.

Uses: The seeds are eaten raw. Heated leaves are pressed firmly over a wound to stop bleeding and are also used to cover fish or meat during cooking. They are reputed to give relief from insect stings. An infusion of the inner bark has been used as a 'wash' for sore eyes.

STRYCHNINE BUSH

Scientific Name:	*Strychnos lucida*
Aboriginal Name:	(Not collected)
References:	87, 102, 113, 118
Habitat:	Mainly amongst rocks, but occasionally in open country, especially on sandstone hills.
Size and Form:	Small tree, 2 – 5 m tall. Dark green foliage. Semi-deciduous. Can be seen with fruits and few leaves at some times of year.
Leaves:	Pear-shaped, up to 7 cm long. Shiny upper surface, dull lower surface, three main veins. Toothed margin when young. Borne at the same level on opposite sides of stems.
Flowers:	White. Tubular, with lobes flaring at mouth to 1 cm wide.
Fruit:	Hard-shelled, globular, 2 cm in diameter. Bright-orange when ripe. Bitter taste.
Uses:	The leaves, berries and roots are boiled in water, and the liquid used to bathe the patient for general illness. This plant must not be taken internally.
Comments:	The fruit and bark contain strychnine and other poisonous substances.

STURT'S DESERT PEA

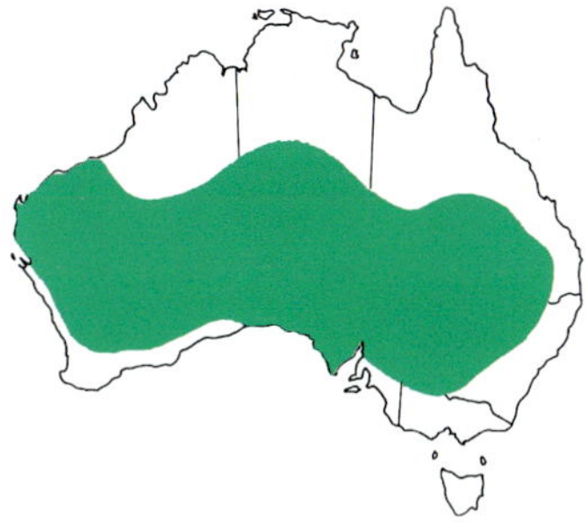

Scientific Name: *Swainsona formosus*

Aboriginal Name: MARLUKURU (Wiluna)

References: 51

Habitat: Widespread in arid regions in sand or sandy loams, especially in disturbed soil along roadsides.

Size and Form: Annual or biennial. Stems growing along ground or bending and growing upwards. Whole plant covered with long soft hairs.

Leaves: Composed of 9 – 17 grey-green oval leaflets, 1 – 3 cm long. Fringed with tiny grey hairs. Borne singly at different heights on the axis.

Flowers: Peaflower-like, 7 – 9 cm long, on erect stems. Petals bright-red, with a black or, rarely, pale pink 'knob' in centre.

Fruit: Hairy pods containing numerous laterally flattened brown seeds.

Uses: Desert people make decorations from the flowers. The seeds are reported to be edible.

Comments: The Aboriginal name refers to the supposed likeness between the flowers and kangaroo eyes. Formerly *Clianthus formosus*.

BLACK LILLYPILLY

Scientific Name:	*Syzygium armstrongii*
Aboriginal Name:	ANGWURRALYA (Kimberley)
References:	2, 24
Habitat:	River banks and dry creek beds.
Size and Form:	Tree, 4 – 20 m tall. Bark fissured, grey and fibrous.
Leaves:	Oval, up to 14 cm long, 4 cm wide. Shiny, slightly paler green on underside.
Flowers:	White, 1.5 cm wide. Resembling small eucalyptus blossoms, but with four distinct petals backing the radiating stamens.
Fruit:	White, apple-like, up to 15 mm long. Some with purple blotches. Turning reddish when ripe.
Uses:	The fruit is edible raw.
Comments:	Tree snakes are often found in these trees, perhaps waiting for small animals that may be attracted by the fruit. There are several other *Syzygium* spp. that bear edible fruits. The exact identification of this tree as *S. armstrongii* is questionable, and a new species may be involved.

RED WILD APPLE

Scientific Name: *Syzygium suborbiculare*

Aboriginal Name: NGARANUDUL (Kalumburu)

References: 77, 92, 113, 126

Habitat: Dry jungle. Sand dunes.

Size and Form: Tree, 6 – 12 m tall depending on habitat.

Leaves: Oval, 20 cm long, 13 cm wide. Thick, tough, shiny. Veins close together.

Flowers: Tubular, up to 2 cm long. Numerous protruding stamens, up to 4.5 cm long. White. Flowering at beginning of wet season, October – November.

Fruit: Round, apple-shaped, 6 cm in diameter. Large round seed loose in central hollow of ripe fruit. Red on outside, flesh white. Ripening December – January or later in wet season.

Uses: The fruit is tart but edible raw or heated and mashed in water with sugar. Some fruit is said to have a hot chilli-like taste. It may be taken to relieve colds. Leaf infusions and fruit pulp are used as topical medicines.

POLYNESIAN ARROWROOT

Scientific Name: *Tacca leontopetaloides*

Aboriginal Name: ADIKALYUBA (Darwin)

References: 2

Habitat: Tropical areas. In sandy soils and as an understorey plant on relict dunes.

Size and Form: Herbaceous perennial arising from a single spherical tuber buried 10 cm below the ground level. Dormant in winter (dry season).

Leaves: Single leaf per plant, deeply indented, 30 cm long. Stem 50 cm long.

Flowers: Up to 18 green drooping flower-heads. Borne on short stalks at top of a stem 1 m tall. Flowering February – March.

Fruit: Flowering head produces several pods, 3 cm long. Pods splitting to reveal globular berries.

Uses: The tubers are potato-like and rich in starch, but bitter and poisonous if untreated. They are coated in eucalypt ash and steamed overnight in an earth oven, then removed, cleaned, coated again and returned to the oven for 36 hours. The berries are reputedly edible.

Comments: The tubers are traditionally gathered in the morning. Meal from the ground tuber is used like arrowroot in Madagascar, India and some Pacific islands.

A SAMPHIRE

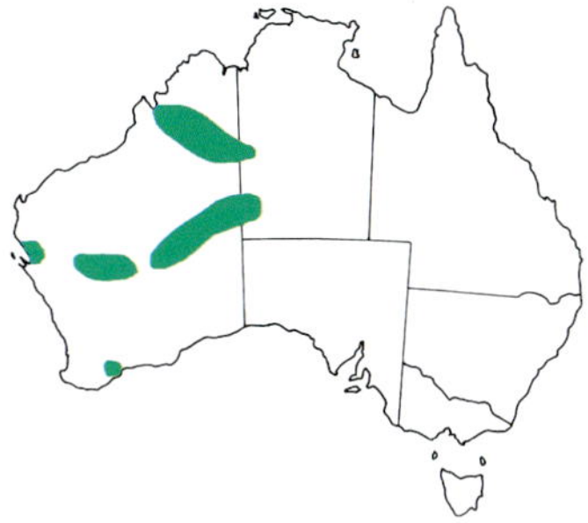

Scientific Name: *Tecticornia verrucosa*

Aboriginal Name: BULLI-BULLI (Murchison)

References: 29

Habitat: Restricted distribution. Only known from a few freshwater claypans.

Size and Form: Succulent shrub, up to 1.5 m tall. Annual or short-lived perennial.

Leaves: Fleshy, stalkless, somewhat obscure structures, closely pressed to stems, 1 – 2 cm long.

Flowers: Insignificant. Fruits borne inside two translucent 'petals'.

Fruit: Fruit wall diaphanous. Seeds brown-black, 1.5 mm long.

Uses: As the tiny seeds ripen, stacks of the plant are made on clear parts of the claypan. As the branches dry out, the seeds fall and can be gathered easily.

Comments: Beside claypans hosting this plant, there are always processing areas and stone grinding slabs. These should not be disturbed, as the interrelationship between such stone artefacts is an important source of archaeological information. A related species, *Tecticornia arborea*, was used similarly.

FISH POISON

Scientific Name: *Tephrosia leptoclada*

Aboriginal Name: (Not collected)

References: 51, 106

Habitat: A variety of habitats, on stony soils and deep sands.

Size and Form: Annual herb, up to 30 cm tall.

Leaves: Composed of 1 – 3 leaflets, occasionally seven. Linear, 20 – 80 mm long, 1 – 2 mm wide. Upper surface hairless, underside covered with short soft erect hairs. Pair of tiny dry membranous appendages at base.

Flowers: Pink, peaflower-shaped, 6 – 7 mm long, short stalks. Widely spaced on long slender flower-heads.

Fruit: Narrow pods, straight or slightly curved, 25 – 35 mm long, 2 – 3 mm wide. Covered with short soft matted hairs. Bean-like seeds, 1.5 mm in diameter.

Uses: The roots are pounded and spread in water holes to 'suffocate' fish and to incapacitate animals that drink the water.

RED CEMENT TREE

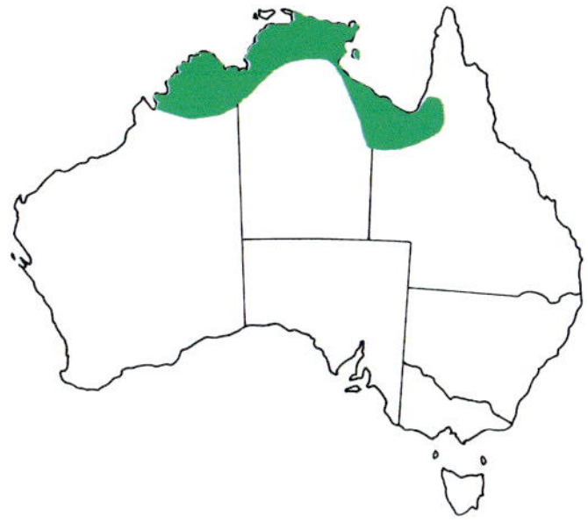

Scientific Name: *Terminalia hadleyana* subsp. *carpentariae*

Aboriginal Name: BUNJAY (Derby)

References: 2, 73, 92, 126

Habitat: Sandstone areas, sandy alluvial plains, stony or shallow well-drained situations.

Size and Form: Small tree, up to 4 m tall. Bark grey, rough but not cracked.

Leaves: Oval, 14 cm long, 10 cm wide. Bright-green, hairy. Deciduous, produced September – October and shed as dry season progresses.

Flowers: Tubular, 3 mm long. Green, stalkless, arranged on spike. Flowering September – October.

Fruit: Almond-shaped, pointed tip, 3 – 4 cm in diameter. Covered with soft hairs, green when ripe. Seed flattened oval, with hard shell surrounded by thin layer of flesh.

Uses: The fruit is eaten raw during the early wet season. Cuts in the trunk cause it to produce clear red gum, which is eaten. Red sap from the inner bark is scraped off, pounded and rubbed on swollen feet as a liniment.

Comments: The bark often shows the scars of cuts made to induce a flow of gum.

BILLY GOAT PLUM

Scientific Name:	*Terminalia latipes* subsp. *psilocarpa*
Aboriginal Name:	GUBINTJ (Derby)
References:	98, 113
Habitat:	Woodlands, along river banks, on a variety of soils.
Size and Form:	Spreading tree, up to 6 m tall.
Leaves:	Oval, 11 cm long, 7 cm wide. Crowded at ends of branches.
Flowers:	Five petal-like lobes surrounding calyx, 10 protruding stamens. White, 3 mm long. Base of calyx tube fused to ovary. Arranged on spike; bisexual flowers at base, male flowers at tip.
Fruit:	Almond-shaped with pointed tip, up to 2 cm long. Purple when ripe, containing a large stone.
Uses:	The pounded bark is used as a poultice on boils, swellings, sores, or to relieve itchiness or ringworm.

WILD PLUM

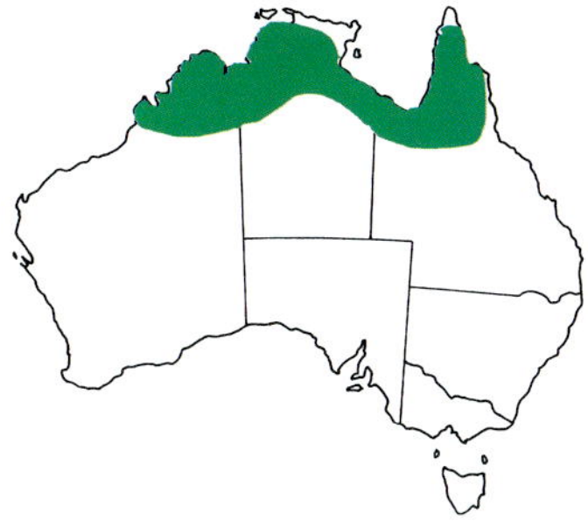

Scientific Name:	*Terminalia platyphylla*
Aboriginal Name:	MANMANGANG (Kununurra)
References:	2, 58, 113
Habitat:	Almost exclusively on creek beds or banks.
Size and Form:	Tree, up to 20 m tall. Bark scaly grey, branchlets furry. Branches often hanging downwards.
Leaves:	Elliptic, narrowing towards base, 7 – 20 cm long, 3 – 13 cm wide. Upper surface almost hairless, furry below.
Flowers:	Tubular, 3 mm long, white. Arranged on spikes up to 17 cm long.
Fruit:	Oblong with a distinct 'beak', 24 mm long, with a thin layer of pithy flesh. Almost hairless when mature.
Uses:	The gum is used as a glue.
Comments:	Suitable for planting in large gardens or in streets.

WINGED-SEED PLUM

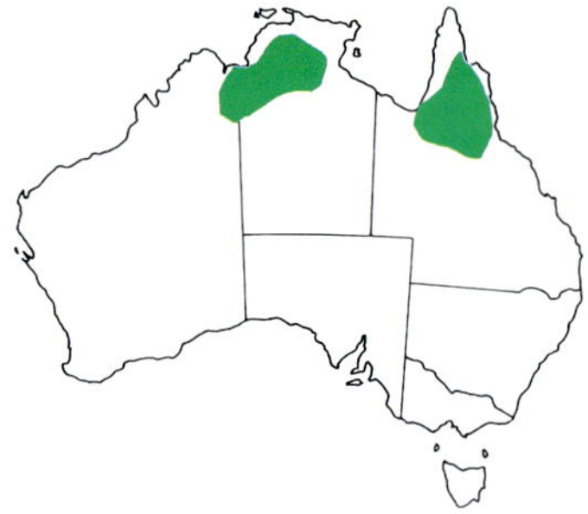

Scientific Name: *Terminalia platyptera*

Aboriginal Name: MABALING (Kununurra)

References: 58, 113

Habitat: Fringes of black soil areas, on river levees. An indicator of clay subsoil on sandy plains.

Size and Form: Usually small tree, occasionally up to 15 m tall. Grey-black deeply cracked bark.

Leaves: Ovate, tapering towards base, 2 – 8 cm long, 1 – 5 cm wide. Darker green above than below, hairy on both surfaces when young. Spirally arranged, crowded on branches.

Flowers: Insignificant, 2 mm long. Arranged in spikes up to 90 mm long, white, densely hairy.

Fruit: Two-winged nuts, 2 – 6 cm long, 4 – 12 cm wide. Green at first, ripening to a straw colour.

Uses: The flowers are a good source of nectar, so native stingless bees often build hives in hollows in these trees. The hives are chopped out of hollows, the honey eaten, and the hard black wax used as a glue or gum in artefact manufacture.

NEW ZEALAND SPINACH

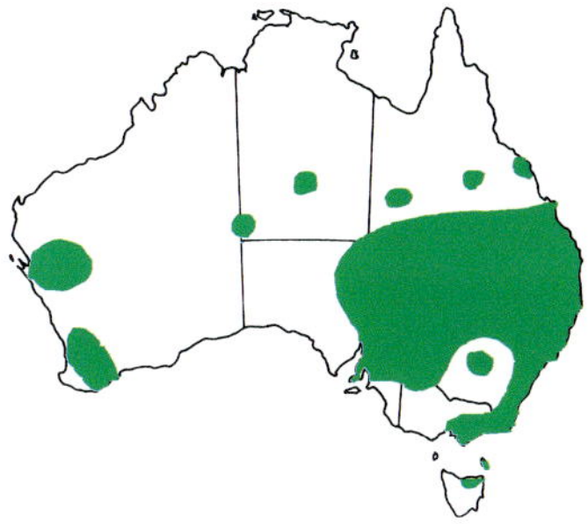

Scientific Name:	*Tetragonia tetragonoides*
Aboriginal Name:	(Not collected)
References:	11, 13, 18, 32, 54, 68, 72, 86, 93, 124
Habitat:	Sandy coastal areas.
Size and Form:	Sprawling fleshy herb. Trailing over dunes or scrambling through low coastal vegetation, 3 m long.
Leaves:	Almost diamond-shaped, 2 – 5 cm long. Thick, bright green. Sparkling because of numerous water-filled pustules on the surface. Stalks to 1 cm long.
Flowers:	Yellow, tubular, up to 8 mm long. Solitary, borne in leaf axils.
Fruit:	Red-black, berry-like, with four longitudinal ribs.
Uses:	The young shoots, leaves and fruits are edible raw or cooked. The leaves taste quite similar to spinach. Boiling in two changes of water renders the bitter-tasting older leaves palatable.
Comments:	Steamed in earth ovens by Aborigines. Widely used 'bush tucker' also called Warrigal Greens.

BEACH YELLOW HIBISCUS

Scientific Name: *Thespesia populneoides*

Aboriginal Name: LURUDA (Broome)

References: 98, 113

Habitat: Tropical shores.

Size and Form: Shrub or tree, up to 5 m tall.

Leaves: Heart-shaped, dark green, 14 cm long, 12 cm wide.

Flowers: Yellow-orange, hibiscus-like, with five petals, 8 cm wide. Solitary.

Fruit: Spherical, slightly flattened, like a tiny pumpkin, 2 cm in diameter. Leathery. Not splitting open at maturity.

Uses: The light wood of vigorous young sappy shoots, which arise from buds on older branches, is used to make the shafts of multi-pronged fishing spears. These float back to the surface after being cast at fish.

Comment: The blossoms are reputedly edible.

CLIMBING FRINGE LILY

Scientific Name:	*Thysanotus patersonii*
Aboriginal Name:	TJUNGOORI (south-west Western Australia)
References:	16, 72, 76, 78, 83
Habitat:	Widespread, ranging from coastal sands and heaths to forest.
Size and Form:	Climbing herbaceous perennial, up to 1.5 m tall. Tubers spindle-shaped, blunt, 1 – 3 cm long, 0.5 cm wide. In clusters of 5 – 10 per plant, radiating out from a compact stem base.
Leaves:	Thread-like, 3 – 6 cm long, glaucous, clustered at base of stem. Often persistent in young plants. Produced infrequently.
Flowers:	Five violet-pink petals, 8 mm long, 4 mm wide, each with three veins. Inner petals broader, oval. Six stamens, three larger than the others. Opening only during daylight.
Fruit:	Globular three-celled capsules, containing numerous angular black seeds.
Uses:	The tubers are eaten raw.
Comments:	The tubers are replaced annually during the growing season.

FRINGED LILY

Scientific Name:	*Thysanotus thyrsoides*
Aboriginal Name:	(Not collected)
References:	13, 71, 72, 83
Habitat:	Mainly in woods or forests on calcareous sands or lateritic soil in jarrah forest.
Size and Form:	Grass-like herbaceous perennial, up to 25 cm tall. Clustered roots expanding into spindle-shaped tubers, 1 – 2 cm long, 3.5 – 5 cm from the stem stock.
Leaves:	Linear, almost cylindrical towards tip, with membranous 'wings' at base, 20 – 30 cm tall. Arising from base, 5 – 15 leaves. Upright, somewhat fleshy, drooping with age.
Flowers:	Purple. Three fringed and three plain tepals. Stalks jointed, erect, 6 – 22 mm long. Flower head 20 – 60 cm tall, three branches, with cylindrical hairless stem.
Fruit:	Three-celled globular capsules of persistent perianth segments. Seeds black with warty seed-coat, 5 mm long, 2.5 mm in diameter.
Uses:	The tubers, which are watery, can be roasted and eaten.
Comments:	The tubers are of highest nutritional value during dormancy.

SNAKE VINE

Scientific Name: *Tinospora smilacina*

Aboriginal Name: NGALYIPI (Balgo)

References: 58, 92, 98

Habitat: Forests and woodlands, winding up support trees.

Size and Form: Woody creeping vine, twisting around trees. A milky latex is exuded from broken twigs.

Leaves: Roughly heart-shaped, up to 16 cm long. Papery to fleshy, with smooth margins. Borne singly at different heights on axis.

Flowers: Green, 2 mm wide, in long spikes in leaf forks. Flowering late autumn to early spring.

Fruit: Oblong berries, red when ripe.

Uses: Used as a pain killer. Stems are cut up, boiled, strained and wrapped around the afflicted area. The liquid is used for treating snakebite or for bathing aching parts.

Comments: This plant is used for making many different things, including straps for baby-carriers. The leaves are used to decorate spears. In former times, broken limbs were straightened then set using a binding of the vine.

JITTA

Scientific Name:	*Tribonanthes australis*
Aboriginal Name:	JITTA (south-west Western Australia)
References:	48, 76, 83
Habitat:	Seasonally wet soils in swamps and around granite rocks.
Size and Form:	Slender herbaceous perennial, up to 20 cm tall. Distinctive, nearly spherical tuber, 1 cm in diameter, buried 1 – 5 cm deep.
Leaves:	Two or three, cylindrical, hollow with sheathing bases, 5 – 8 cm long.
Flowers:	Single or multi-flowered, on stems 10 – 30 cm tall. Petals 6 – 11, white, woolly, linear, tapering towards tip. Six stamens, filaments slightly flattened, anthers linear.
Fruit:	Ovoid capsules. Woolly, leathery, splitting at tip to release numerous grey-brown striated seeds.
Uses:	The tuber, which is replaced annually, is eaten.

CAMEL BUSH

Scientific Name: *Trichodesma zeylanicum*

Aboriginal Name: PAWALARRI (Roebourne)

References: 13, 76, 78, 83, 93

Habitat: Widely distributed, mainly in sandy or gravelly soils in open country.

Size and Form: Coarse, hairy herb, up to 1 m tall.

Leaves: Oblong linear to slightly tapered, 7 – 10 cm long. Borne singly at different heights on the stem, or at the same height on opposite sides of the stem.

Flowers: Blue. Five broad petals with narrowly pointed tips. Small leaf-like bract under each flower stalk.

Fruit: Four single-seeded shining nutlets attached by their whole inner face to a stem, surrounded by the calyx.

Uses: A decoction of the plant is rubbed on sores.

Comments: In India it is considered to be diuretic, and is also used as a cure for snakebite. There are at least 2 sub-species in Australia.

SOFT SPINIFEX

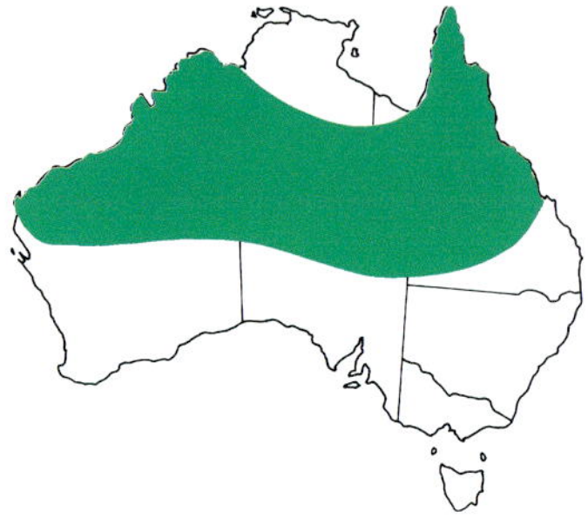

Scientific Name:	*Triodia pungens*
Aboriginal Name:	JINALPA (Balgo)
References:	16, 51, 61
Habitat:	Beside creek and river channels.
Size and Form:	Straggly tussock, 30 cm tall, 1.2 m wide. Resinous, sticky. Some trailing stems produce roots at the stem joints. Rarely ring-forming.
Leaves:	Narrow, spiky, highly resinous. Soft, not prickly.
Flowers:	Flowering stems erect, up to 90 cm tall. Typical grass-like terminal many-flowered spikes.
Fruit:	Small seeds held in the flower parts.
Uses:	Whole plants are used to make low windbreaks for sleeping shelters. Gum exuded from the leaves is gathered into balls, after being threshed from the plants and warmed to make it sticky. Used to haft spearheads and chisel blades. 'Wax' is also collected from *Triodia plectrachnoides.*

BUCK SPINIFEX

Scientific Name:	*Triodia* aff. *wiseana*
Aboriginal Name:	PARU (Roebourne)
References:	None found.
Habitat:	Open plains and hill slopes. Widespread.
Size and Form:	Hummock grass with rigid sharp points, up to 1 m wide.
Leaves:	Linear, usually rolled into a spike to prevent transpiration.
Flowers:	Grass-type flower stalks, rising up to 1 m above the plant. Flowers seed quickly. Flowering mid to late summer if sufficient rainfall.
Fruit:	Seeds in flower stalks. Like grains of millet, 2 mm long.
Uses:	The seeds are ground and mixed with water. This dough is then cooked in clean hot sand to produce a nutritious 'damper'.
Comments:	Seeds can be dug out of storage chambers in the nests of certain species of ant. Clumps of some types of spinifex are pounded between two stones until fibrous, like cotton wool, and then spun into a string.

CREEK HAZEL

Scientific Name:	*Trymalium floribundum*
Aboriginal Name:	(Not collected)
References:	71
Habitat:	Sheltered parts of jarrah and karri forests.
Size and Form:	Tall single-trunked shrub, up to 5 m tall. Often growing in thick groves. Bark smooth and blackish.
Leaves:	Roughly oval, 8 cm long, 5 cm wide. Dense white hairs on underside. On short stems. Young leaves bright pale-green. Older leaves distinctly paler on underside.
Flowers:	White, tiny, in clusters towards the ends of branches.
Fruit:	Tiny insignificant capsules.
Uses:	The leaves provide a 'bush soap'. Slender trunks are used by Aborigines in south-west Western Australia to make shaved-stick ornaments.
Comments:	Several other species occur in Australia, and can probably be used in a similar way.

BULL RUSH

Scientific Name: *Typha domingensis*

Aboriginal Name: YANJET (Perth)

References: 13, 57, 67, 72, 83, 93, 127

Habitat: Banks and shallows of pools and semi-permanent waterholes. Not usually in flowing water.

Size and Form: Crowded erect leaf bunches, often topped by characteristic flower spikes, up to 2 m tall.

Leaves: Strap-like, 2 m long, 3 cm wide, arising from submerged or buried rhizome.

Flowers: Tall slender stalk. Flowers and seeds contained within cinnamon-brown cylindrical velvety structure, 10 – 15 cm long, up to 2 cm diameter.

Fruit: Tiny seeds held in same structure as flowers.

Uses: The leaf bases of the young shoots and horizontal roots are stringy, but they can be eaten after roasting and pounding. The stems are used for weaving, thatching and string-making. The dry fluffy flower parts are used in fire-lighting and pillow stuffing. Both the tiny seeds and the masses of pollen are edible, either steamed or used as coarse flour

Comments: Many references are made to the use of the rhizome. Sometimes it was peeled, roasted, pounded to remove fibrous strands and then baked into a loaf. Newly emerging shoots are eaten 'like asparagus'.

SMALL NETTLE

Scientific Name: *Urtica urens*

Aboriginal Name: (Not collected)

References: 18, 86, 93

Habitat: Found only in association with habitation (because introduced), particularly in disturbed damp areas.

Size and Form: Annual herb. Seldom exceeding 50 cm tall. Stems and leaves with stinging hairs.

Leaves: Oval, up to 5 cm long.

Flowers: Male and female flowers occur grouped together. Calyx of male flowers 1 mm in diameter. Female flowers 1 mm long.

Fruit: Nut enclosed in flower, 2 mm long.

Uses: The leaves and young stems are edible. They are boiled for a few minutes to remove the formic acid, then used as a vegetable.

Comments: A strong twine is made from the stem fibres of some nettle species in Europe and America. The plant has square purple stems. The stinging hairs on the stems and leaves of this herb give an infallible, if uncomfortable, guide to its identity.

SUPPLEJACK

Scientific Name: *Ventilago viminalis*

Aboriginal Name: WALAKARI (Balgo)

References: 58, 61, 84, 85, 91, 102

Habitat: On banks, depressions or flats adjacent to watercourses.

Size and Form: Erect medium-sized tree, 3 – 10 m tall. Young branches flexible and drooping.

Leaves: Narrow, tapering towards tip, 5 – 15 cm long, 0.6 – 1.2 cm wide. Prominent light-coloured mid-vein. Short stalk. Borne singly at different heights on the branches.

Flowers: Flower heads 2 – 7 cm long, borne in angle between leaf and stem. Composed of many greenish to cream-yellow flowers, 5 mm in diameter. Lacking petals, but with five spreading sepals.

Fruit: Globular nut, 4 mm diameter, at one end of an oval wing, 20 – 40 mm long. Wing thin, leathery, green-yellow.

Uses: The hard wood is used for making boomerangs. A decoction of the roots and bark is used as a liniment to relieve aches and pains, and as a mouthwash.

Comments: Gum exuded from cuts in the bark is reputedly edible and sweet.

PENCIL YAM

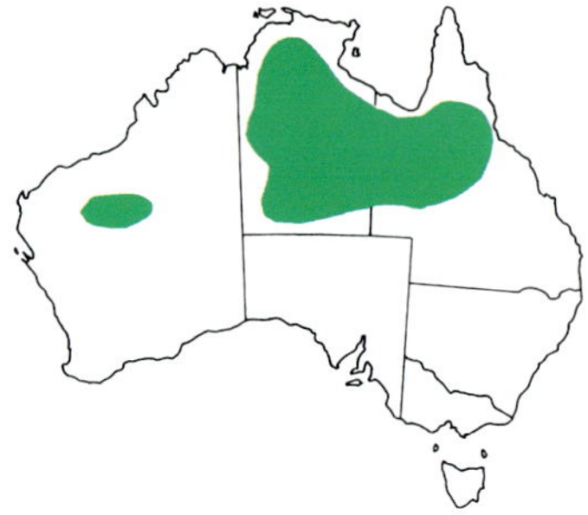

Scientific Name:	*Vigna lanceolata*
Aboriginal Name:	TJIRILPATJA (Balgo)
References:	3, 13, 47, 54, 62, 102.
Habitat:	Usually in dry sites along watercourses. Sometimes on sandplains with spinifex.
Size and Form:	Perennial slender twining herb, with prostrate stems to 2 m long. Sparsely covered with hairs.
Leaves:	Composed of three broad ovate leaflets, tapering towards the tip, 2 – 8 cm long. Borne on a main stalk, the terminal leaflet some distance along the stalk from the others.
Flowers:	Yellowish, 'peaflower'-like, 7 mm long. Borne in clusters at ends of stems 7.5 – 20 cm long, which arise from leaf axils.
Fruit:	Somewhat turgid legumes, slightly sickle-shaped, pointed, 30 – 45 mm long, 4 – 6 mm wide. Breaking open spirally.
Uses:	The long thin roots are roasted on the fire or in hot ashes, then peeled before eating.
Comments:	The roots are reputed to be eaten raw in central Australia.

NATIVE BROOM

Scientific Name:	*Viminaria juncea*
Aboriginal Name:	KOWEDA (Perth)
References:	18, 76, 83, 93
Habitat:	Usually near swamps or on damp soils around lakes.
Size and Form:	Erect shrub, 2 – 4 m tall. Long wiry pendulous branches.
Leaves:	Rarely found. If present: narrow, ovate, 45 mm long, in groups of two or three leaflets.
Flowers:	Yellow, 'peaflower'-shaped, 7 – 10 mm wide. Arranged in long, often pendulous, flower-heads.
Fruit:	Brown or black pods, 5 mm long. Containing smooth brown seeds, 4 mm long.
Uses:	Used in conjunction with bark for thatching shelters.
Comments:	The leaves of *Xanthorrhoea preissii* were also used for thatching (see entry).

BLACK GRAPE

Scientific Name:	*Vitex glabrata*
Aboriginal Name:	MIDJERAN (Kununurra)
References:	2, 12, 58, 102, 120, 126
Habitat:	Sandy soils, near base of rocky hills and in gullies.
Size and Form:	Tree, 4 – 20 m tall.
Leaves:	Composed of three leaflets, rarely five, on stems up to 3 cm long. Each leaflet 12 cm long, 7 cm wide. Hairy when young, dark green, shiny.
Flowers:	Tubular, up to 11 mm long, with one prominent and four lesser petal-like lobes. White with purple hairs. Protruding stamens. Arranged in many-flowered sprays, up to 5 cm long.
Fruit:	Globular berries, up to 1 cm in diameter, black when ripe.
Uses:	The sweet-tasting fruit is eaten raw. The wood is used for firesticks. A good shade tree.
Comments:	In the Northern Territory, the crushed fruits are relished by the elderly, as they require no chewing. They are also used in ceremonies.

BLACKBOY

Scientific Name: *Xanthorrhoea preissii*

Aboriginal Name: BALGA (Perth)

References: 47, 72, 76, 83, 102

Habitat: Widespread and common in a variety of habitats, especially low-lying damp areas.

Size and Form: Monocotyledon resembling a tree, up to 4 m tall. Stem distinctly fire-blackened, composed of old compacted leaf bases around a vascular core.

Leaves: Numerous, linear, rhomboidal in cross-section, 50 – 110 cm long, pointed. Rigid, brittle, borne in dense spreading tuft at top of stem.

Flowers: White persistent perianth, composed of six unfused segments, with six stamens. Numerous, crowded on upper two-thirds of flowering stem.

Fruit: Three-celled ovary. Mature capsules conical, hard, smooth, 1 – 1.2 cm long. Protruding from the spike, with a pointed tip.

Uses: The flowering stem yields a gum that is used for food. Gum from the trunk makes a cement used in artefact manufacture. The bases of the young leaves are edible. Dry flower stems are used to make fire by friction.

Comments: More than 20 uses are recorded for *Xanthorrhoea preissii*. Aboriginal names for three other *Xanthorrhoea* spp. are known for the Perth region, but it is now impossible to determine which scientific name should be ascribed to these, and consequently, which uses were made of them.

GLOSSARY

alkaloid: any of a group of nitrogenous basic compounds found in plants. Common examples are morphine, strychnine and quinine.

annual: a plant that completes its life cycle in one year.

annular: in the form of a ring.

apical: at the apex or highest point.

axil: the angle between a stem and a leaf.

axillary: borne in an axil.

biennial: a plant that completes its life cycle in two years.

bipinnate: having compound leaves in which the leaflets are arranged in two rows, one each side of a midrib.

bisexual: having both fertile male and female organs in the same flower.

bract: a leaf-like structure that surrounds or encloses a flower or group of small individual flowers.

bracteole: a small leaf-like structure borne on the stalk or calyx of a flower.

breakaway: a small gully or low cliff line formed by erosion.

bulb: a storage organ consisting of a short, usually underground, stem surrounded by swollen leaf-bases.

calyx: the outer series of leaves in a flower, each one a sepal.

capsule: a dry fruit that opens when mature to release the seeds.

carpel: the structure that bears and encloses the ovules in flowering plants. Normally comprises the ovary, a stigma and usually a style.

corm: a fleshy swollen stem base, usually underground, in which food reserves are stored between growing seasons.

corolla: the whole of the petals of a single flower.

corona: a ring of tissue seated upon the corolla.

corymb: flower structure in which each flower has a separate stem, those of the lower flowers being longer than those of the flowers above, resulting in a flat-topped cluster of flowers.

damper: an unleavened loaf or scone, typically cooked in an open fire.

deciduous: falling or shedding of any plant part at the end of the growth period.

decoction: the solution resulting from boiling a medicinal plant in water.

dicotyledon: a flowering plant whose embryo has two seed leaves (cotyledons).

dioecious: having the male and female reproductive structures on separate plants (cf. *monoecious*).

drupe: a succulent fruit with a stone enclosing the seed or seeds. Peaches and plums are examples.

ellipsoid: (three dimensional shape) ellipse shaped in plan and elevation.

elliptic: oval in outline, widest at the centre.

epiphyte: a plant that grows on another plant, but not as a parasite.

eviscerate: to remove the internal organs.

family: a group of genera that are botanically related.

fissured: a surface covered with small cracks or splits.

floret: a small flower. In grasses, the small inconspicuous flower and the bracts enclosing it.

gall: an abnormal outgrowth caused by certain parasitic insects, fungi, bacteria or mechanical injury.

genus: a natural group of closely related species (plural *genera*)

gland: a structure, within or on the surface of a plant, with a secretory function (adjective *glandular*).

glaucous: covered with bloom, giving a greyish or powdery appearance.

globose: nearly spherical.

habitat: the environment within which a plant grows.

heath: low growing many branched woody shrub.

herb: plant without woody stems (adjective *herbaceous*)

hoary: densely covered with short whitish hairs.

inrolled: sides (of a leaf) rolled inwards to the central rib.

keel: a ridge, like the keel of a boat (adjective *keeled*)

kwongan: an Aboriginal word for heathland and mallee heath areas.

lateral: fixed on or near the sides of an organ.

laterite: a residual soil rich in iron and aluminium hydroxides.

lenticular: shaped like a biconvex lens.

liana: a climbing or twining plant.

linear: long and narrow with more or less parallel sides.

lobe: a rounded division of leaf, petal or other plant organ.

mallee: a shrub with many stems arising from a large woody sub-surface structure bearing dormant buds (usually applied only to eucalypts).

margin: the edge of a leaf.

microspore: the smaller (male) of the two kinds of spores produced in the sexual life cycle of a heterosporous plant.

monocotyledon: a flowering plant whose embryo has only one seed leaf (cotyledon) (cf. *dicotyledon*).

monoecious: male and female flowers on the same plant (cf. *dioecious*)

mucilage: slimy material exuded by certain plants or plant organs (adjective *mucilaginous*).

mulga: common name for a group of acacias, and sometimes a scrub composed of a dense growth of acacias.

nut: a hard dry fruit that does not open at maturity, formed from two or more carpels but containing only one seed.

obloid: (a three-dimensional shape) with short parallel sides and rounded ends, as if composed of two hemispheres linked together by a very short cylinder.

oblong: having the length greater than the width but not many times greater, and the sides parallel.

opposed: pairs (of leaves) at equal heights on a stem.

opposite: arising at the same level but on opposite sides.

ovary: the basal portion of a carpel or group of fused carpels, enclosing the ovule(s).

ovate: shaped like a section through the long axis of an egg, with the broadest part at the base.

ovoid: egg-shaped (in three dimensions).

panicle: an irregular, much-branched spray of flowers.

peaflower: flower structure with parts arranged in two vertical planes (see sketch).

pedicel: the stalk of a flower.

pendulous: hanging downwards.

perennial: a plant with a life span of more than two years.

perianth: the outer part of a flower, consisting of fused calyx and corolla.

persistent: remaining attached to the plant beyond the expected time of falling (e.g. of sepals not falling after flowering).

petal: one of the, usually conspicuous, segments forming the inner whorl of the perianth.

petiole: the stalk portion of a leaf.

phyllode: a flattened leaf stalk that functions as a leaf.

pinna(e): a compound leaf, divided once with leaflets arranged on each side of a central stalk (adjective *pinnate*).

pinnule: a leaflet of a bipinnate leaf.

pistil: a free carpel or a group of free carpels.

prostrate: lying close to the ground.

raceme: an unbranched spray of equally stalked flowers along a common stem (adjective *racemose*).

receptacle: the axis of a flower.

rhizome: a creeping underground stem.

rhomboid: roughly diamond shaped.

samphire: fleshy leafed salt tolerant plant characteristic of salt-marsh communities.

scape: a flowering stem.

sepal: one of the, usually green, segments forming the outer of the two whorls of leaf-like structures of a flower, and collectively known as the calyx.

serrated: notched on the edge, with asymmetrical teeth pointing forward.

shrub: a woody plant less than five metres tall, either without a distinct main axis, or with branches persisting on the main axis almost to its base.

smoking: a practice of traditional Aboriginal medicine involving building a fire rich in coals, smothering it with numerous branches of the selected plant and immersing the patient in the ensuing thick white smoke.

spadix: a spicate inflorescence, with a stout often succulent axis.

spathe: a large leafy bract enclosing a flower cluster.

species (*sp.* or *spp.*): a population of living organisms that is distinguished from others by recognisable features, and that can grow together with closely related species with little or no hybridisation occurring (plural *species*).spike: an arrangement of stalkless flowers along an individual stem (adjective *spicate*).

sporangium: a structure within which spores are formed (plural *sporangia*).

spore: a unicellular or few-celled asexual or sexual reproductive unit, not containing an embryo.

sporocarp: a fruiting body containing sporangia.

stamen: one of the male organs of a flower, which consists of a pollen-bearing anther and a supporting filament or stalk.

striated: striped with parallel longitudinal lines or ridges.

stigma: the receptive tip of a carpel.

style: the elongated part of a carpel or group of fused carpels, between the ovary and the stigma.

subtend: to stand below or close to.

succulent: juicy or fleshy, applied to fruits, leaves or stem.

taproot: the main, descending root of a plant that has a single dominant root axis.

tendril: a twisting thread-like shoot of a leaf or stem by which a climbing plant may hold itself in position.

tepal: a perianth segment in a flower in which all the perianth segments are similar in appearance.

terminal: at the end or apex.

touchwood: decayed wood or punk used as tinder.

truncate: with an abruptly transverse end, as if cut off.

tuber: the swollen end of an underground stem (adjective *tuberous*).

valve: the segment of a capsule that splits to release the seeds.

vascular: *of plants*, containing tissues specialised for the conduction of fluids.

vegetative: the leaves, rhizomes and other non-reproductive parts of a plant.

vein: a strand of vascular tissue.

REFERENCES

1. Aboriginal Communities of the Northern Territory (1993). 'Traditional Aboriginal Medicines in the Northern Territory of Australia.' (Conservation Commission of the Northern Territory: Darwin.)
2. Altman, J.C. (1984). The dietary utilisation of flora and fauna by contemporary hunter-gatherers at Momega Outstation, North Central Arnhem Land. *Aust. Aborig. Studies* **1**, 35—46.
3. Anderson, R.H. (1968). 'The Trees of New South Wales.' (New South Wales Government Printer: Sydney.)
4. Aston, H.I. (1973). 'Aquatic Plants of Australia.' (Melbourne University Press: Melbourne.)
5. Basedow, H. (1918). Expedition of exploration in North West Australia. *Transact. Royal Geog. Soc. Australasia* **XVIII**, session 1916—1917.
6. Beard, J.S. (1970). 'A Descriptive Catalogue of West Australian Plants.' (Society for Growing Australian Native Plants and Kings Park Board: Perth.)
7. Bennett, E.M. (1988). 'The Bushland Plants of Kings Park, Western Australia.' (Kings Park Board: Perth.)
8. Bird, C., and Beeck, C. (1988). Traditional plant foods in the south-west of Western Australia: the evidence for salvage ethnography. *in*: B. Meehan and R. Jones 'Archaeology with Ethnography: an Australian Perspective.' (Dept of Prehistory, Research School of Pacific Studies, Australian National University: Canberra.)
9. Blackall, W.E., and Grieve, B.J. (1981). 'How to Know Western Australian Wildflowers' Part IIIB. (University of Western Australia Press: Nedlands.)
10. Blundel, V.J. (1975). 'Aboriginal Adaptation in Northwest Australia.' PhD Thesis, University of Wisconsin (unpubl.).
11. Bolam, A.G. (1927). 'The Trans-Australian Wonderland.' 6th Ed. (A.G. Bolam: Melbourne.)
12. Boomsma, C.D. (1981). 'Native Trees of South Australia.' *Bulletin* **19**, 2nd Ed. (Woods and Forests Dept: Adelaide.)
13. Bourne, G.M. (1953). The food of the Australian Aboriginal. *Proc. Nutr. Soc.* **12**, 58—65.
14. Brand-Miller, J., James, K.W., Maggiore, P.M.A. (1993). 'Tables of Composition of Australian Aboriginal Foods.' (Aboriginal Studies Press: Canberra.)
15. Brittan, N.H. (1981). Revision of the genus *Thysanotus* R.Br. (Liliaceae). *Brunonia* **4**, 67—181.

16. Brokensha, P. (1975). 'The Pitjantjatjara and their Crafts.' (Aboriginal Arts Board of Australia Council: North Sydney.)

17. Bureau of Flora and Fauna (1984 and continuing). 'Flora of Australia.' (Australian Government Publishing Service: Canberra.)

18. Chauncey, P. (1878). Notes and anecdotes of the Aborigines of Australia. pp 221—284 *in*: R.B. Smyth (ed.) 'The Aborigines of Victoria.' (Victorian Government Printer: Melbourne.)

19. Chippendale, G.M. (1973). 'Eucalypts of the Western Australian Goldfields and the Adjacent Wheatbelt.' (Dept of Primary Industry: Canberra.)

20. Chippendale, G.M. (1981). 'Eucalyptus Buds and Fruits.' (Australian Government Publishing Service: Canberra.)

21. Cleland, J.B. (1936). Ethno-botany in relation to the Central Australian Aborigine. *Mankind* **2**(1), 6—9.

22. Coon, C.S. (1972). 'The Hunting People.' (Jonathon Cape: London.)

23. Craig, G.F. (1983). 'Pilbara Coastal Flora.' (Western Australian Dept of Agriculture: South Perth.)

24. Crawford, I.M. (1982). 'Traditional Aboriginal Plant Resources in the Kalumburu Area: Aspects in Ethno-economics.' (Western Australian Museum: Perth.)

25. Cribb, A.B., and Cribb J.W. (1975). 'Wild Food in Australia.' (Fontana/Collins: Sydney.)

26. Cribb, A.B., and Cribb J.W. (1981). 'Wild Medicine in Australia.' (Collins: Sydney.)

27. Cribb, A.B., and Cribb, J.W. (1982). 'Useful Wild Plants in Australia.' (Fontana/Collins: Sydney.)

28. Cunningham, G.M., Mulham, W.E., Milthorpe, P.L., and Leigh, J.H. (1992). 'Plants of Western New South Wales.' (Inkata: Melbourne.)

29. Dix, W.C., and Lofgren, M.E. (1974). Kurumi: possible incipient agriculture associated with a stone arrangement. *Rec. West. Aust. Mus.* **3**(1), 73—78.

30. Elliot, W.R., and Jones, D.L. (1982). 'Encyclopaedia of Australian Plants Suitable for Cultivation.' Vols 1—6. (Lothian: Melbourne.)

31. Erickson, R., George, A.S., Marchant, N.G., and Morcomb, M.K. (1973). 'Flowers and Plants of Western Australia.' (Reed: Sydney.)

32. Everist, S.L. (1981). 'Poisonous Plants of Australia.' Rev. Ed. (Angus & Robertson: Sydney.)

33. Frankel, D. (1982). An account of Aboriginal use of the yam-daisy. *Artefact* **7**(1 & 2), 43—45.

34. Gardner, C.A. (1975). 'The Manna Wattle (*Acacia microbotrya* Benth).' *Bulletin* 2433. (Department of Agriculture: Perth.)

35. Gardner, C.A. (1979). 'Eucalyptus of WA.' *Bulletin* 4013. (Department of Agriculture: Perth.)

36. Gardner, C.A., and Bennets, H.W. (1956). 'The Toxic Plants of WA.' (West Australian Newspapers Ltd: Perth.)

37. George, A.S., (n.d.). 'The Banksia Book.' (Society for Growing Australian Plants, NSW.)

38. Goddard, C., and Kalotas, A. (eds) (1985). 'Punu, Yankunytjatjara Plant Use.' (Angus & Robertson: North Ryde.)

39. Gott, B. (1982). Ecology of root use by the Aborigines of Southern Australia. *Archaeol. Oceania* **17**, 59—67.

40. Gott, B. (1982). *Kunzea pomifera* - Dawson's 'Nurt'. *Artefact* **7**, 13—17.

41. Gott, B. (1983). *Microseris scapigera*: a study of a staple food of Victorian Aborigines. *Aust. Aborig. Studies* **2**, 2—18.

42. Gould, R.A. (1969). 'Yiwara: Foragers of the Australian desert.' (Collins: London.)

43. Griffin, G. (1985). Bushtucker. *Aust. Plants* **13**, 127—134.

44. Hall, N., Johnston, R.D., and Chippendale, G.M. (1975). 'Forest Trees of Australia.' (Australian Dept of Agriculture, Forestry and Timber Bureau: Canberra.)

45. Harmer, J. (1985). 'North Australian Plants, Part 1: Top End Wildflowers.' (Surrey Beatty: Chipping Norton.)

46. Harris, T.Y. (1966). 'Wild Flowers of Australia.' (Angus & Robertson: Sydney.)

47. Hassell, E. (1975). 'My Dusky Friends: Aboriginal Life, Customs and Legends and Glimpses of Station Life at Jarramungup in the 1880s.' (C.W. Hassell: Fremantle.)

48. Hiddins, Capt. L.J. (1980). 'Survive to Live' (Government Printer: Canberra.)

49. Holliday, I., and Watton, G. (1977). 'A Field Guide to Banksia.' (Rigby: Adelaide.)

50. Isaacs, J. (1987). 'Bush Food.' (Weldons: McMahons Point.)

51. Jessop, J. (ed.) (1981). 'Flora of Central Australia.' (Australian Systematic Botany Society and Reed Books: Sydney.)

52. Jones, D. (1984). 'Palms in Australia.' (Reed: Frenchs Forest.)

53. Jones, D.L., and Clemsha, S.C. (1980). 'Australian Ferns and Fern Allies.' 2nd Ed. (Reed: Sydney.)

54. Jones, D.L., and Gray, B. (1977). 'Australian Climbing Plants - with Notes on their Cultivation.' (Reed: Sydney.)

55. Kelly, S. (1969, 1978). 'Eucalypts.' Vols 1&2. (Thomas Nelson: Melbourne.)

56. Kenneally, K.F. (1977). 'The Natural History of the Wongan Hills, Handbook II.' (Western Australian Naturalists' Club: Perth.)

57. King, P. (1986). 'Plant Identikit.' (Conservation Commission Northern Territory: Darwin.)

58. Kofod, F. (n.d.). 'List of Mirriwung Plant Names: Kununurra region.' (Archaeology Dept, Western Australian Museum.) (unpubl.)

59. Lane-Poole, C.E. (1919). 'A Forest Policy for Australia'. (Institute of Science and Industry, Government Printer: Melbourne.)

60. Lassak, T.V., and McCarthy, T. (1983). 'Australian Medicinal Plants.' (Methuen: North Ryde.)

61. Latz, P.K. (1995). 'Bushfires and Bushtucker; Aboriginal plant use in central Australia.' (IAD: Alice Springs.)

62. Latz, P., and Griffin, G.F. (1978). Changes in Aboriginal land management in relation to fire and to food plants in Central Australia. pp 77—85 *in*: B.S. Hetzel, and H.J. Frith 'The Nutrition of Aborigines in Relation to the Ecosystems of Central Australia.' (CSIRO: Melbourne.)

63. Lazarides, M., and Hince, B. (1993). 'CSIRO Handbook of Economic Plants of Australia.' (CSIRO: Melbourne.)

64. Lear, R., and Turner, T. (1977). 'Mangroves of Australia.' (University of Queensland Press: St Lucia.)

65. Levitt, D. (1981). 'Plants and People: Aboriginal Use of Plants on Groote Eylandt.' (Australian Institute of Aboriginal and Torres Strait Islander Studies: Canberra.)

66. Lindsay, H.A. (1967). 'The Bushman's Handbook.' (Jacaranda: Hong Kong.)

67. Low, T. (1985). 'Wild Herbs of Australia and New Zealand.' (Angus & Robertson: North Ryde.)

68. Maiden, J.H. (1975). 'The Useful Native Plants of Australia.' (Compendium: Melbourne.)

69. Maiden, J.H. (1892). Vegetable exudations. *Trans. Roy. Soc. S.A.* **16**(1), 1—9.

70. Manson Bailey, F., and Gordon, P.R. (1887). Plants reputed poisonous and injurious to stock. *Qld Agric. J.* **11**(2), 1—4. (Queensland Dept of Agriculture: Brisbane.)

71. Marchant, N.G., Wheeler, J.R., Rye, B.L., Bennett, E.M., Lander, N.S., and Macfarlane, T.D. (1987). 'Flora of the Perth Region.' Vols 1&2. (Dept of Agriculture: Perth.)

72. Meagher, S. (1974). The food resources of the Aborigines of the south-west of Western Australia. *Rec. West. Aust. Mus.* **3**(1), 14—65.

73. Meehan, B., Gaffey, P., and Jones, R. (1979). Fire to steel: Aboriginal exploitation of Pandanus and some wider implications. pp 73—80 *in*: P. Lauer (ed.) 'Readings in Material Culture.' *Occas. pap. in Anthrop.* **9**. (Anthropology Museum, Queensland University: Brisbane.)

74. Melville, G. (1947). An investigation of the drought pastures of the Murchison district of Western Australia. *J. Dept Agric. West. Aust.* **24**(1), 1—29.

75. Mitchell, A.A., and Wilcox, D.G. (1988). 'Arid Shrubland Plants of Western Australia.' (University of Western Australia Press: Nedlands.)

76. Moore, G.F. (1974). 'Diary of Ten Years Eventful Life of an Early Settler in Western Australia and also a Descriptive Vocabulary of the Language of the Aborigines.' (University of Western Australia Press: Nedlands.)

77. Moore, D.R. (1979). 'Islanders and Aborigines at Cape York.' (Australian Institute of Aboriginal and Torres Straits Islanders Studies: Canberra.)

78. Mullins, B., and Baglin, D. (1978). 'Western Australian Wildflowers in Colour.' (Reed: Sydney.)

79. Nind, I.S. (1979). Description of the natives of King George's Sound. pp 15—55 *in*: N. Green (ed.) 'Nyungar, the People.' (Creative Research and Mt. Lawley College: Perth.)

80. Paddy, E. & S., and Smith, M. (1993). 'Boonyja Bardag Gorna: All Trees are Good for Something.' 2nd Ed. Community Report 87/1. (Western Australian Museum: Perth.)

81. Palmer, A. (1985). 'Yalgoo.' (LAP Industries: Fremantle.)

82. Parker, A. (1980). An ethnobotany of the Western Desert, Leonora WA, 1977. *Newsletters* **13**, 37—42. (Australian Institute of Aboriginal and Torres Straits Islander Studies: Canberra.)

83. Pate, J.S., and Dixon, K.W. (1982). 'Tuberous, Cormous and Bulbous Plants.' (University of Western Australia Press: Nedlands.)

84. Piele, A.R. (1976). Gugadja Ethnobotany. *in*: Papers presented at the Ethnobotany Workshop Biennial Conference. (Australian Institute of Aboriginal and Torres Straits Islanders Studies: Canberra.)

85. Piele, A.R. (1980). Preliminary notes on the ethnobotany of the Gugadja Aborigines of Balgo, WA. *West. Aust. Herb. Res. Notes* **3**, 59—64.

86. Poiner, G. (1976). The process of the year among Aborigines of the central and south coast of New South Wales. *Arch. Phys. Anthrop. Oceania* **11**(3), 186—206.

87. Petheram, R.J., and Kok, B.L. (1983). 'Plants of the Kimberley Region of Western Australia.' (University of Western Australia Press: Nedlands.)

88. Randell, B.R. (1992). Mulga: a revision of the major species. *J. Adel. Bot. Gard.* **14**, 105—132.

89. Reid, E. (1977). 'The Records of Western Australian Plants used by Aborigines as Medicinal Agents.' (Western Australian Institute of Technology: Bentley.) (unpubl.)

90. Richmond, G.S. (1993). A review of the use of *Eremophila* (Myoporaceae) by Australian Aborigines. *J. Adel. Bot. Gard.* **15**(2), 101—107.

91. Roth, W.E. (1903). Notes on savage life in the early days of West Australian settlement. *Proc. Roy. Soc. Qld* **17**, 45—69.

92. Scarlet, N., White, N. and Reid, J. (1982). Bush medicines - the pharmacopoea of the Yolngu of Arnhem Land. *in*: J. Reid (ed.) 'Body, Land and Spirit.' (University of Queensland Press: St Lucia.)

93. Scott, M.P. (1972). Some Aboriginal food plants of the Ashburton District, WA. *West. Aust. Naturalist* **12**(4), 94—96.

94. Semeniuk, V., Kenneally, K.F., and Wilson, P.G. (1978). 'Mangroves of Western Australia.' (Western Australian Naturalists' Club: Perth.)

95. Simmons, M. (1981). 'Acacias of Australia.' (Nelson: Melbourne.)

96. Simon, B.K. (1978). A preliminary checklist of Australian Grasses. *Tech. Bull.* **3** (Botany Branch, Dept Prim. Industries: Brisbane.)

97. Sked, J. (1980). 'Bush Medicine.' (Pine Rivers Society for Growing Australian Plants: Lawnton.)

98. Smith, M., and Kalotas, A.C. (1985). Bardi plants. *Rec. West. Aust. Mus.* **12**(3), 317—359.

99. Smith, N., Wididburu, B., Nuwallat Harrington, R., and Wightman, G. (1993). Ngarinyman ethnobotany: Aboriginal plant use from the Victoria River area, Northern Australia. *North. Terr. Bot. Bull.* **16**. (Conservation Commission of the Northern Territory: Palmerston.)

100. Smith, N.M., and Wightman, G.M., (1990). Ethnobotanical notes from Belyuen, Northern Territory Australia. *North. Terr. Bot. Bull.* **10**. (Conservation Commission of the Northern Territory: Palmerston.)

101. Smith, M.V. (1993). 'Recherche a L'Esperance: a Prehistory of the Esperance Region of South-western Australia.' PhD Thesis, University of Western Australia, Dept of Anthropology (unpubl.).

102. Storman, E.J. (1977). 'The Salvado Memoirs.' (University of Western Australia Press: Nedlands.)

103. Sullivan, S. (1978). Aboriginal diet and food gathering methods in the Richmond and Tweed River valleys, as seen in early settler records. *in*: I. McBryde (ed.) 'Records of Times Past, Ethnohistorical Essays on the Culture and Ecology of the New England Tribes.' (Australian Institute of Aboriginal and Torres Straits Islanders Studies: Canberra.)

104. Symon, D.E. (1979). Fruit diversity and dispersal in *Solanum* in Australia. *J. Adel. Bot. Gard.* **I**, 321—331.

105. Thompson, D.F. (1936). Notes on some bone and stone implements from North Queensland. *J. Royal Anthrop. Inst.* **LXVI**, 71—78.

106. Thompson, D.F. (1949). 'Economic Structure and Ceremonial Exchange Cycle in Arnhem Land.' (Macmillan: Melbourne.)

107. Tindale, N.B. (1972). The Pitjandjara. pp 217—268, *in*: M.G. Bicchieri (ed.) 'Hunters and Gatherers Today.' (Holt, Rinehart & Winston: New York.)

108. Turner, J. (1986). Indigenous and ingenious: Aboriginal plant usage. *Landscope* **1**(4), 10—15.

109. Tyler, J.P. (1988). 'The Dampier Salt Guide to Pilbara Plants for the Garden.' (Dampier Salt Pty Ltd: Karratha.)

110. Webb, L.J. (1948). Guide to the medicinal and poisonous plants of Queensland. *Bulletin* 232. (CSIRO: Melbourne.)

111. Webb, J.L. (1969). The use of plant medicines and poisons by Australian Aborigines. *Mankind* **7**, 127—146.

112. Webb, L.J. (1975). Eat, die and learn, the botany of the Australian Aborigines. *Identity* October, 2—11.

113. Wheeler, J.R. (ed.), Rye, B.L., Koch, B.L., and Wilson, A.J.G. (1992). 'Flora of the Kimberley Region.' (Dept of Conservation and Land Management: Perth.)

114. Whibley, D.J.E. (1980). 'Acacias of South Australia.' (South Australian Government Printer: Adelaide.)

115. Wightman, G., and Mills, L. (1991). 'Bush Medicine Identikit.' (Conservation Commission Northern Territory: Darwin.)

116. Wightman, G., and Andrews, M. (1991). 'Bush Tucker Identikit.' (Conservation Commission Northern Territory: Darwin.)

117. Wightman, G., Dixon, D., Williams, L., and Dalywaters, I. (1992). Mudburra Ethnobotany: Aboriginal plant use from Kulumindini (Elliot), Northern Australia. *North. Terr. Bot. Bull.* **14**. (Conservation Commission of the Northern Territory: Palmerston.)

118. Wightman, G., Jackson, D.M., and Williams, L.L.V. (1991). Alawa Ethnobotany: Aboriginal plant use from Minyerri, Northern Australia. *North. Terr. Bot. Bull.* **11**. (Conservation Commission of the Northern Territory: Palmerston.)

119. Wightman, G.M., and Smith, N.M. (1989). Ethnobotany, vegetation and floristics of Milingimbi, Northern Australia. *North. Terr. Bot. Bull.* **6**. (Conservation Commission of the Northern Territory: Palmerston.)

120. Wightman, G.M., Garalnganjak Roberts, J., and Williams, L. (1992). Mangarrayi Ethnobotany: Aboriginal plant use from the Elsey area, Northern Australia. *North. Terr. Bot. Bull.* **15**. (Conservation Commission of the Northern Territory: Palmerston.)

121. Wightman, G. M. (1989). Mangroves of the Northern Territory. *North. Terr. Bot. Bull.* **7**. (Conservation Commission of the Northern Territory: Palmerston.)

122. Wilson, K.L. (1981). Revision of the genus *Mesomelaena* (Cyperaceae). *Telopea* **2**, 181—197.

123. Winfield, C. (1985). 'Bush Tucker: a guide to, and Resources on Traditional Aboriginal Foods of the North West of South Australia and Central Australia.' (Wattle Park Teachers Centre: Wattle Park.)

124. Wontolla (n.d.). 'Bushcraft: How to Live in Jungle and Bush.' (F.H. Johnston: Sydney.)

125. Woodward, P.(1986). 'An Australian Herbal.' (Hyland House: Melbourne.)

126. Worsley, P. (1961). The utilisation of natural food resources by an Australian Aboriginal tribe. *Acta Ethnographica Academiae Scientiarum Hungaicae* **10**, 153—190.

127. Wright, B.L. (1970). 'Notes and Tapescripts on Some Plants of the Pilbara and their Traditional use by the Aborigines.' (Dept of Aboriginal Sites, Western Australian Museum: Perth.) (unpubl.)

128. Zola, N., and Gott, B. (1992). 'Koorie Plants Koorie People.' (Koorie Heritage Trust: Melbourne.)

INDEX

NOTES

NOTES